Ministry of Agriculture, Fisheries & Food

STRAWBERRIES

by

HILARY M HUGHES

ADAS Regional Fruit Specialist
MAFF, Burghill Road
Westbury on Trym, Bristol BS10 6NJ

Reference Book 95

ondon: Her Majesty's Stationery Office

First published 1937
Ninth edition 1980

Any reference to proprietary products in this book should not be construed as an official endorsement of those products, nor is any criticism implied of similar products which are not mentioned.

0 11 240336 0

Foreword

There have been many changes in strawberry crop production over the past few years so this Book, replacing the previous Bulletin 95, has been completely rewritten and edited by the original author, under the sponsorship of the National Fruit Group of the Agricultural Development and Advisory Service.

The section on insect pests was revised by Mr H J Gould and that on diseases by Messrs D Wiggell, W T Dale and Dr M B Simkin. Mr P A T Steer wrote the section on mechanical harvesting and Mr A G Jones that on herbicides. Mr J B Williams assisted with overall editing.

This publication, together with leaflets published more frequently by the Ministry, offers growers, farmers, students and keen gardeners an authoritative account of all aspects of production of this queen of English fruits.

D J Fuller
Senior Horticultural
Advisory Officer

Ministry of Agriculture, Fisheries and Food

August 1979

Contents

PLATES

General Considerations

The maincrop strawberry, using up-to-date methods of propagation, varieties and weed control, is a comparatively easy crop to grow on reasonably fertile soils. Frosts in May can damage the flowers, and so a frost-free site is essential. Sheltered sites encourage growth and crop. Soil drainage must be adequate or various root troubles may occur. Although a wide rotation is advisable and will reduce problems from soil-borne pests, diseases and weeds, many strawberries are grown under little or no rotation to take advantage of favourable sites. As a result serious soil disease problems can occur.

The major problems now are harvesting and marketing. A hectare of strawberries yielding 7–12 tonnes over a three to four week period requires at least 24 full day pickers, plus supervision. The picking is laborious and not all labour is either willing or able to do the job. The fruit is extremely perishable and must be kept picked regularly. Wet weather can create serious problems. Because of these problems there has been a large increase in the number of holdings where the whole crop is sold to the public through pick-your-own (PYO).

Over the last few years steady progress has been made in the mechanical harvesting of strawberries and recent progress is particularly promising. This method of harvesting, if fully successful, will be of particular interest to growers producing fruit on a large scale for processing outlets. It is not yet certain whether mechanically harvested fruit will be suitable for the retail trade.

Strawberries offer a quicker return on capital outlay than any other fruit since, under special methods of cultivation, a crop can be picked as early as the first summer after planting. The fruit is the first of the season's home-grown supplies to reach the markets and, while the principal demand is for dessert, it is also used for jam-making, canning and quick-freezing.

The total area grown has changed frequently in the past. Since statistics were first collected it has fluctuated from 10 800 ha in 1935 to the low level of 3680 ha under war-time restrictions in 1943. From 1966 to 1979 the area varied between 8040 and 6000 ha. More careful growing, the provision of improved stocks, the use of herbicides and more efficient sprays for the control of pests and diseases have led to higher yields, with the result that total production has increased from a smaller area. Over the ten years 1968 to 1978 the average annual area was 7000 ha and the average total annual production 47 000 tonnes. In 1978 the estimated output from 5700 ha was nearly £34 million.

The main growing region is in eastern England followed by the south-east and he west-midlands, with smaller areas elsewhere. The area grown outside the raditional areas is increasing because of the development of pick-your-own arms.

In 1977 owing to the drought years of 1975 and 1976 the total area grown was less than usual. The main growing areas in 1977 were:

	ha
Kent	1046
Norfolk	985
Hereford and Worcester	400
Cambridgeshire	390
Essex	319
Lincolnshire	306

followed by Humberside, Cheshire, Staffordshire, Hampshire and Warwickshire.

The areas in Norfolk, Cambridgeshire, Lincolnshire and Humberside are mainly concerned with production for processing and are adjacent to the factories, which also process other horticultural produce grown in these areas. A large proportion of the Kent and Essex crop is sold as fresh fruit because of proximity to the London markets. Depending on the prices and markets, a proportion of the crop in the other major producing counties is sold as fresh fruit direct to local shops, or through pick-your-own, farm shops, or through wholesale commission markets.

Large areas of strawberries on any one holding are exceptional; this crop tends to be produced by many growers from quite small units in comparison with other horticultural crops. Very small units of only a few, or even one, hectare are usual in the areas where early strawberries are produced. Some growers relying on pick-your-own sales grow 20–30 ha as do those growing for processing, but more usually this crop is grown in 2–5 ha blocks. This fragmentation indicates small scale production, often on a family basis, and the difficulty of coping with picking on a large scale.

In the major strawberry growing areas there is competition for suitable labour. Elsewhere female casual labour may be more readily available, but needs training and is unlikely to pick as quickly. Thus choice of a suitable site and provision of adequate labour are two essentials for successful growing, plus easy access to markets or processors, or the ability to sell all the fruit through pick-your-own sales.

There has been some development in farm pre-cooling of fruit and in refrigerated transport, but generally fruit picked one day must be sold and consumed the next. Strawberries cannot satisfactorily be re-consigned from one market wholesaler to another, nor do they store successfully unless cooled and then only for a few days. Because consignments from individual growers are usually small, many wholesalers arrange to collect fruit in the main growing areas. Some growers have also arranged co-operative transport by road, or occasionally by rail. Thus in non-traditional strawberry growing areas the advantage of labour availability may be offset by marketing problems. However isolated producers of strawberries can often sell all their produce through pick-your-own sales, or farm shops, or direct to local shops and hotels.

The production of early strawberries, grown under protection and ripe in May and early June, is concentrated in the counties of the west, south and south-east where warm spring and autumn weather aid early growth, flowering and ripening.

Various methods for protecting and forwarding the crop are used. There is much site variation and early strawberries can be produced on warm sheltered sites elsewhere. Early strawberries generally meet a good demand but recently the expense incurred in the production of this fruit has not always been recouped through adequate returns, particularly when yields per plant are low. It is doubtful whether the production of forced strawberries from heated glasshouses is economic. Market prices in April are affected by imports of strawberries from the USA, Europe and elsewhere, which are now readily available by air freight. Prices for forced strawberries may not always give a sufficient return for the use of heated glass in comparison with other crops.

Late strawberries from colder areas in the northern counties, or from elevated or north facing fields, often meet a ready sale in August, when the main production period of late June and July is over. A recent development has been the production of autumn strawberries from varieties that produce second crops, or from perpetual varieties, and demand is satisfactory at this time.

It was partly because of the high cost of harvesting this fruit and its very perishable nature that the pick-your-own sales started and have been so popular and beneficial to growers and customers alike. As a result many farmers are growing strawberries only for pick-your-own sales, particularly those well situated relative to the main urban areas and the road network. This has resulted in a larger area of the crop being grown away from traditional areas. Other established fruit growers may market some of their crop through this outlet. Even in the specialist early areas pick-your-own customers are brought in for the later ripening fruit, after the early, high value fruit has been sent to market. In addition to pick-your-own, more and more growers who are suitably situated may sell the picked fruit in punnets from farm shops and roadside stalls. It is difficult to estimate the proportion of the crop that is now sold direct on the farm, but it may well be over half the crop that is grown for the retail market, excluding that grown primarily for processing.

Growth of the Strawberry

The cultivated strawberry belongs to the genus *Fragaria* in which there are about 12 species, some native to Europe and Asia, others to North and South America. The native wild strawberry, found in England along hedge banks and woods is *F. ananassa vesca* the wood or alpine strawberry, a small plant with tiny leaves and flowers and very small, dark scarlet, pippy fruit. This species was cultivated in gardens for many years, together with the 'hautbois' *F. elatior*, whose fruits are very little larger. The hautbois strawberry, unlike *F. ananassa vesca*, has separate male and female plants (dioecious), with the stamens produced only on the flowers of one plant and the berry and seed bearing part on separate plants: these need interplanting to ensure good crops.

During the early eighteenth century species of strawberry plants were brought to Europe from the new lands of North and South America. In eastern America, *F. virginiana* the hardy meadow strawberry grows abundantly, and along the western coasts of North and South America the less hardy, large fruited beach strawberry, *F. chiloensis* is found. It appears that all modern hybrid, large-fruited strawberries have originated from the genetical admixture arising from the crosses between these two American species. Some of the original crosses would have been by chance, but English and French nurserymen at the beginning of the nineteenth century were breeding strawberries to obtain better fruit size and soon achieved varieties such as Keen's Seedling and Black Prince which, although now no longer grown, apparently did not differ much in fruit size or flavour and plant appearance from present day varieties. It is because of the hybrid (heterozygous) nature of the modern strawberry that such a wide range of flower, fruit and plant characteristics are seen. This has enabled many local varieties to be evolved throughout the world, each suitable for local conditions. During this century, plant breeding became a more exact science and the raising of new strawberry varieties has been carried out chiefly at research institutes using plants obtained from other breeders, particularly from America.

The Plant

The strawberry is a low-growing herbaceous plant with a short perennial crown or rootstock on which the leaves are produced close together. Branch crowns and runners are produced in the axils of the leaves which are normally trifoliate. The fruit trusses are produced terminally on the crown but appear to arise laterally because they are displaced by new crown growth before they emerge. The flower trusses may each have few, or many, flowers on branching stalks.

The main roots arise from the rootstock and, together with numerous side roots, form a fibrous mass. As the plant becomes older, main roots are frequently produced rather high on the rootstock and the crowns protrude 5–8 cm out of the soil. In Britain the main scaffold roots are chiefly produced during the period after fruiting until late autumn when soil temperatures become too low for further growth. In the spring the growth of lateral roots is more abundant. The roots of healthy well-grown strawberries in good soil can extend to a considerable depth and range and their growth is extremely rapid.

The strawberry plant will continue root and crown growth throughout the year if conditions are suitable, but in most parts of Britain growth ceases during the autumn when the plant becomes dormant and recommences as the air and soil temperatures increase in the spring. The younger leaves remain green in the winter although the older leaves gradually die. In wet or severe winters there is noticeably more winter killing of the foliage than during mild winters. In some parts of the world the winter temperatures are so low that the crowns, unless protected by a straw mulch, can be killed, but this rarely occurs in England. Growth of the strawberry plant is much earlier in the south-west and south of England where mild, late autumns and warm early springs ensure a much longer period of growth than in areas with longer winters.

The strawberry plant grows rapidly in the spring after winter chilling, unfolding new leaves and producing flower trusses. In the field the plants are in flower during May and June depending on variety and season but earlier if protected by cloches or tunnels. The plants are in flower for at least four weeks. This long flowering period is due to the fact that the trusses emerge in sequence on each crown and the flowers open in sequence on each truss. Older plants with many crowns have a longer flowering period than plants (runners) formed the previous year. These one-year-old plants usually flower before the older plants and have only a small number of trusses with fewer flowers on each truss, and thus produce earlier fruit and generally bigger berries. Older plants produce heavier crops per plant, but of smaller berries, because of the competition for plant nutrients between the numerous berries on the one plant.

Each flower of the strawberry nominally has ten green sepals and five white petals enclosing 30–40 stamens and numerous carpels on the receptacle but flowers with more sepals and petals are frequently observed.

The flower trusses are initiated in the crown of each plant during the previous autumn. To a great extent this determines the yield the following year, since the crop is determined by the number of berries. Individual berry size, for each variety, is determined by the number of fertilised seeds as well as cultural factors. An older plant with numerous branch crowns and good food reserves in roots and the rootstock generally initiates a plentiful supply of flower trusses. A one-year-old plant, planted the previous autumn, winter or spring may only have one or two trusses, or may be 'blind' with no trusses. All blind plants will initiate trusses as usual the following autumn. The number of trusses on young plants is controlled by runner age and size, date of planting, weather and variety.

The Fruit

The fruit, or strawberry, is not a true fruit in the botanical sense, since it consists of the swollen receptacle on the surface of which are embedded the numerous achenes or seeds.

Self pollination is easily effected in the strawberry because the stamens carrying the pollen sacs are arranged around the stigmas of the carpels within the flower. Although insects and hive bees are seen on strawberries it is probable that wind movement and moisture may also help in pollination. In some varieties, the earliest flowers to open often have aborted anthers (Plate III) and therefore no viable pollen and the pollen grains of other anthers may also be non-viable. This may be one of the main causes of the malformation of the earliest fruits, particularly under protected cultivation.

To produce a well shaped berry it is necessary for a high proportion of carpels to be pollinated and fertilised; if some are not fertilised there is no hormone stimulation from the seed (achene) into the adjacent portion of the receptacle which, consequently, does not grow. Strawberry varieties differ considerably in the number of seeds per berry, in their size and colour, and whether they are embedded, or stand out on the surface of the fruit.

There are also big variations in the size of the calyx (or cap) and varieties with large leafy sepals are undesirable as they detract from the bright colour of the fruit when picked. Each individual flower and berry is attached to the main flower truss by a thinner stalk. For ease of picking it is desirable that this stalk should be readily severed and some varieties break off much more easily than others. There are also marked differences between the attachment of the berry to the calyx; some varieties pull away easily, others with difficulty, and in others the plug may pull out leaving the fruit rather damaged. This latter point is important in varieties that are to be used for processing where the green calyx has to be removed.

After the flower is fertilised it is generally about four to six weeks before the berry is ripe. Early varieties have a shorter maturation period and most early varieties tend to be lighter in crop than mid-season varieties.

Natural Vegetative Propagation

The strawberry increases by the production of runner plants on stolons, as well as through seed. Seed is not a suitable method of propagation for cultivated varieties as they do not come true to type. The stolon, a creeping stalk, is produced in the leaf axil and grows some way out from the parent plant during the summer. At the second node a runner plant is formed and a new stolon arises on the runner plant to continue the runner train, which often branches. The runner plant produces roots, at first a few anchoring roots and then a more extensive fibrous root system. Thus the strawberry naturally propagates itself by this vegetative method of runner production (Plate I). Once the runner plant is big enough the stolon may be severed from the mother plant and the runner plant removed and planted elsewhere. Many stolons may be formed on a mother plant and, as many runner plants can arise from one stolon train which branches to form numerous runners and stolons, a carpet of several hundred of these daughter plants may be formed in one season.

Because most varieties of strawberries runner freely in summer under suitable conditions of day-length and temperature, each variety can be readily propagated and its characteristics maintained. All the plants of any one variety have thus been derived from the single, original, new seedling by this method of vegetative propagation; all are exactly the same as the original plant and are collectively called a clone.

Under cultivated conditions in fruiting plantations the runners produced by the original plants may be entirely removed by cultural treatments and the original plants grown on for two, three or four years as spaced plants. Alternatively all, or preferably some, of the runner plants are allowed to root along the rows, excess runner plants being removed. The rows thus consist of rooted runners and the crowns of the original plants forming what is known as a matted row. This gives a high plant population and high yields under suitable soil conditions, but can result in overcrowding of the crowns, disease and small fruit. Sometimes the original plant crowns die or are crowded out.

Although fruiting beds will persist for very many years by this means of natural vegetative propagation, most commercial plantations have to be replanted usually every three to six years because of weeds, overcrowded rows, virus or other disease problems. Virus diseases (page 97) were responsible for the disappearance of many of the older strawberry varieties, but nowadays the certification schemes maintain stocks successfully.

Response of the Strawberry to Temperature and Day-length

Research work has shown that the temperature and day-length (photo-periodism) have a considerable effect on the growth of the strawberry, probably through the control of the production of plant hormones. The varieties grown in Britain start to initiate and produce runners during the lengthening summer days. Flower trusses are initiated during the period August–November, to bloom in the following year.

Removal of the leaves after fruiting (defoliation) is a common cultural practice and provided this is done within a week of the end of harvest does not reduce the crop. In Scotland and northern England, with some varieties, defoliation may in fact increase the following crop by promoting flower formation.

Because of the response of varieties to climatic conditions, particularly of day-length and temperature, it is not uncommon to find varieties behaving very differently in different geographical regions. Northern European varieties lose vigour and may fail to runner, but flower profusely and crop poorly when grown in warmer parts of the world nearer the equator where they are never exposed to the necessary long-day conditions and low chilling temperatures.

Although most strawberry varieties grown in Britain produce only one crop per year, this is not so with all varieties. Redgauntlet and some other varieties, if protected so flowering is earlier in the spring, will initiate further flower trusses during spring to early summer that will produce fruit in the autumn, and also initiate flower trusses for the normal crop the following year. In very hot and dry summers even unprotected plants may behave in this manner though the fruit seldom ripens satisfactorily in the autumn. Under plastic film houses Redgauntlet flowers and crops almost perpetually from spring to autumn. In New Zealand in the open Redgauntlet will flower and crop over six months. Certain varieties grown in California also behave in a similar manner producing about four flushes of fruit: in Britain they only produce a single crop.

The so-called perpetual, or everbearing, strawberry varieties (page 22) initiate and produce flowers throughout the growing season and some produce flowers on the newly formed runner plants even before these are properly rooted. Most of the perpetual varieties have the disadvantage that, at any one time, there is no large quantity of fruit ripe and, although they may flower and fruit from late June until the November frosts, the total yield may not be greater than from a single-crop variety. These varieties may be of use on pick-your-own holdings.

It may be possible to breed single crop varieties to extend the fruiting season beyond the present normal period of mid-June to the end of July. There is a demand for late varieties to extend the season and help avoid gluts of fruit.

The great adaptability of the strawberry, the different methods of controlling plant density, growth and crop, the variation between varieties and the constant battle to maintain weed, pest and disease free plants, provide great interest and sometimes reward for the grower prepared to study this interesting crop.

Propagation and the Need for Healthy Plants

As has been described the strawberry is propagated by planting runner plants, which are produced on stolons from established plants during the summer (page 6), and this method of vegetative propagation perpetuates all the characteristics of the parent.

Many experiments have shown that virus disease infection reduces vigour and yield to an appreciable extent, even when no obvious symptoms are produced. It is essential to plant virus tested, or as healthy material as possible.

Most virus diseases are transmitted from infected to healthy strawberry plants by aphids (greenfly), but some are transmitted by soil-inhabiting eelworms, and leaf hoppers. The aphid-transmitted virus diseases are of most widespread importance. Virus diseases are in the sap of the plant, and runners from infected parent plants, although often apparently healthy, will be similarly infected. There is no cure for virus infection. Thus the provision of virus-free runners is of prime importance to the strawberry industry. The availability of virus-tested plants, through the certification and nuclear schemes described, is one of the reasons for the improved vigour and yield obtained from present day varieties.

Tested Clonal Stocks

It is very important to propagate from clonal stocks of which the history, health standard and cropping capacity are known. (A clone is a race of plants propagated vegetatively from a single plant). The first testing and selection of strawberries was begun at East Malling Research Station in 1931, and since then tested clones of all the more important varieties have been regularly produced there. When necessary, specially treated and retested clonal stock of existing varieties is produced from time to time, and virus tested clones of new varieties are made available to the Nuclear Stock Association as requested, for propagation and sale.

Certification Schemes for Runner Plants

In order that growers may buy runners of a known standard of health, the Ministry of Agriculture, Fisheries and Food (MAFF) has for many years arranged for inspection of plants and the award of certificates to those plants attaining the required standards.*

---

*The condition relating to certification and the varieties and stocks available under the schemes are liable to alteration from time to time. Particulars of current conditions and an annual Register of growers of certified strawberry plants are available, free, from the Ministry of Agriculture, Fisheries and Food, Plant Health Administration Unit, Eagle House, 90–96 Cannon Street, London EC4 N6HT. Tel.: 01-623 4266. The Register is issued as soon as possible after the final inspections have been carried out, usually in November.

These schemes have raised the general vigour and health standard of strawberry plants and growers should only use certified plants. A copy of the certificate, or the certificate number, should always be obtained when the plants are purchased.

There are two types of certificates, SS (Special Stock) and A, normally available to the grower. Plants bearing these are derived from Nuclear Foundation and Elite Stocks handled through the Nuclear Stock Association. Plants of A certificate are suitable for planting for fruiting, but should not be used if SS certified runners are obtainable at the right time. SS plants, because of the special conditions of raising, afford the grower extremely healthy material.

Any grower wishing to plant runners early each autumn would be well advised to use SS plants each year from which to propagate his own runners. Runners can be taken from one-year-old fruiting plants but this tends to exhaust the plant and reduce the crop in the first fruiting year: a special runner bed, preferably isolated from the fruiting beds, for early runners is therefore recommended.

Nuclear, Foundation and Elite Stock Schemes

New clones are raised and existing clones are distributed through the Nuclear Stock Scheme. Plants of approved clones are released by East Malling Research Station as mother plants (Nucleus) to the Nuclear Stock Association Ltd., which has its registered office at Agriculture House, 25–31 Knightsbridge, London SW1X 7NJ. This was formed as a growers' co-operative organisation during 1953 for the purpose of:

> the propagation of any nuclear horticultural stock, including seeds, which may be required by raisers of Special Stock grown in accordance with standards laid down by the Ministry of Agriculture, Fisheries and Food (MAFF), and
>
> the sale of nuclear horticultural stock to eligible raisers of Special Stock.

The scheme ensures that these stocks are owned by an independent body, and at the same time that close and continuous technical supervision is provided by MAFF until the stocks are ready for distribution to the Special Stock (SS) growers.

The nuclear stock is grown in strict isolation and under special cultural conditions, usually in gauze houses, structures covered with fine mesh nylon gauze which exclude insects. The Association arranges the propagation of these plants on a contract basis with a few selected growers. The resultant crop of runners is carefully inspected and if found to attain the required standard of health is certified as Foundation Stock.

This Foundation Stock is distributed on contract to selected growers who are prepared to comply with the conditions laid down for propagation. Provided that the runners from these plantings attain the required health standards, they are certified as Elite Stock, which is sold by the Association to runner-raisers for producing SS strawberry plants.

Special Stock (SS) Certificate

The propagation system developed at East Malling Research Station now forms the basis of the requirements for the Special Stock Certificate. The essential features are:

1. runner-raising is treated as a special undertaking entirely separate from fruit production. No land scheduled as infected with red core (page 93) can be used for the production of certified runners; strawberries must not have been grown on the land during the previous two years, and the site must be approved before planting.
2. runner nurseries are established well away from existing strawberry areas: 1.5 km from any plantation not entered, or eligible for, SS and A certification and 200 m from any private garden with strawberries.
3. runners are propagated on the isolated block or isolated strip system, to allow systematic roguing and removal of any diseased plants and runners in blocks during inspections. Different varieties must not be grown within 4 m of each other. The plants are generally de-blossomed.
4. only virus-tested clonal material approved by the Ministry for the purposes of the scheme, and normally only stocks that obtained an Elite Certificate in the previous year, are eligible.

The runner beds are inspected twice during the growing season and at the final inspection the plants must be a 100 per cent true to type and of a high standard of health.

SS certificates are not issued for any stock found to be infected with red core disease, or appreciably affected with mildew or strawberry aphids, or with any other pest or disease, or lacking in vigour.

Stock Certificate A

Runners intended for the A Certificate are subject to less stringent standards than for the SS. Mother plants of SS stocks of approved varieties have to be grown to show the normal vigour and habit of the variety and must be not less than 50 m from any plant obviously infected by serious pest or disease. Plants entered must not be grown on land on which strawberry plants were grown at any time during the previous year, or on land scheduled with red core disease. There are no restrictions on plant spacing; plants may be fruited and the runners allowed to root in the alleys. Adjoining beds of different varieties must be separated by at least 3 m and the intervening space kept free from runners. Two inspections are made during the growing season and the A Certificate is issued for stocks that have been found to be at least 99.5 per cent true to type and substantially free from visible diseases and pests.

Isolation and Choice of Site

Although strawberries will grow well on most soils, heavy and tenacious soils are not suitable for runner-raising. Runners have to be lifted in the autumn and winter and, if much soil clings to the plant, the lifting becomes both difficult and expensive. Where irrigation is available very light sandy soils produce good runners but irrigation is essential. Peat soils produce very well rooted runners but weed control is often a problem. Runner beds can be situated on fields liable to spring frosts as damage to the flowers does not matter.

There are very few conditions for A stock plants, and these have already been mentioned. For raising runners for Special Stock certification, isolation is more important and the requirements have been listed. These isolation requirements are designed to reduce any virus infection by strawberry aphids. The wild strawberry plant (*Fragaria ananassa vesca*), however, does not constitute a danger. Verticillium wilt is a disease that can be carried in the soil or in infected runner plants. The disease is widespread and erratic in occurrence (page 94). Land known to be infected should not be used for runner production.

Soil-borne virus diseases which are spread by certain soil eelworms (page 98) have recently been recognised and are fairly widespread. Soil samples should be taken to see if the soil is free from contamination. To avoid infestation by leaf or stem eelworms the rotation must not include certain host crops such as beans (page 88).

Any grower who is contemplating raising plants under the SS scheme for the first time should get in touch with the local ADAS Horticultural Adviser in order that the scheme can be fully explained and advice given on the high standards of cultivation and control of pests and diseases required and the site and necessary isolation approved.

Planting the Runner Bed

There are two systems of plant raising, the isolated block system in units of not more than six plants, or the isolated strip system. In the isolated block system planting is carried out so that runners can be trained into separate blocks, leaving cultivated alleys between the blocks. If any plant showing pest, disease or virus symptoms is found at any inspection, all the plants in the same block can be removed, without coming into contact with the neighbouring blocks.

The distance required between the plants depends on the variety, the type of soil and growth expected. Spacing varies from 1.2×1.2 m (6900 plants/ha) to 1.8×1.8 m (3100 plants/ha).

The first runners are laid along the rows to allow cultivations to continue, but runners produced later are first trained within the block and allowed to spread beyond the parent plants if necessary, but they must not be allowed to reach into the adjoining blocks.

In the isolated strip system, plants are set out in continuous rows down the field, plants 0.4 to 0.6 m apart in rows 3 to 3.6 m apart. The plants are trained by switching all runners in one direction, at right angles to the row. One advantage of this method is that mechanical cultivation can be maintained ahead of the runners, thereby providing a weed-free friable soil to encourage the runners to root. If any diseased plant is found, a stretch of plants on each side of the diseased plant is removed.

Cultural Operations

The appropriate certified runners for propagation are not usually available for planting until late autumn. However, it is now permissible to plant cold stored runners certified the previous year; these are particularly useful if early runners are required (page 14). Experiments have shown cold stored mother plants will produce earlier and more runner plants than freshly dug mother plants.

Freshly dug runners can be planted in early winter or the spring. Cold stored runners must not be planted before 1 June or after mid-August: July is preferable in order to get them well established. The flower trusses produced a few weeks after planting should be removed if the plants are not growing vigorously.

Some runners will appear in the autumn and must be switched into position, preferably within the row of parent plants.

In the spring all plants may need firming after frost, and it is recommended that all flower trusses be removed to encourage the production of plenty of runners. As the runners develop they are laid into position according to the system adopted. Runners can be anchored with stones or soil. Loose, friable soil well supplied with moisture, or irrigated if necessary, aids rooting. It is important to lay runners in their final positions before rooting starts, as disturbance thereafter results in a poor root system. The process of laying-in must be repeated several times. If an isolated block becomes filled with runners, later ones can be laid round the edges, provided that the isolation between the blocks is maintained.

Weeds must be controlled on most soils by a combination of cultivations in the early part of the year and chemical herbicides applied during June, July or August (page 49). On highly organic soils where herbicides are not always satisfactory, much hand weeding may be necessary. Weeds that are allowed to develop will cause weak, drawn-up runners and make inspection difficult. A good cover of runners will help to smother weeds.

Pest and Weed Control

Careful inspections are necessary, particularly for aphids and mildew. All runner beds should be given routine sprays against aphids in the spring and summer (page 86), and if mildew (page 95), red spider mites (page 86) or strawberry mites (page 87), are noticed, further spraying will be necessary.

Lifting and Marketing Runners

Lifting to supply runners for early districts and for protected cropping begins in August. The main lifting of runners is during late September and October. Lifting continues, when soil conditions allow, through the winter months until March and April when there is demand for runners for spring planting.

In August, runners are forked up and unrooted ones sometimes bedded back if possible. Irrigation may be helpful, both to lift runners under dry conditions and to re-root runners that have been disturbed.

Strict grading and rapid delivery for planting is necessary. Early lifted runners must be kept well covered and damped down and then packed for despatch in moist moss, woodwool or peat to avoid drying out.

Plants can be lifted in three ways: hand forking and bundling in the field; mechanical lifting with hand bundling in the field; or mechanical lifting and placing runners into trays and bundling in the packing shed.

Hand lifting, by casual labour, under piece-rate working is common practice. This rate includes trimming of the stolons, grading and binding into bundles of 25 using rubber bands size 90 × 2 mm.

Mechanical lifting, using potato lifters of the hoover or undercutting types, is being attempted by several plant raisers. Plants are then easily bundled in the field or carted in trays to a packing shed. Where several varieties are grown in a field great care must be taken to avoid mixing the varieties.

The yield of runners under either the isolated block or isolated row system varies from 124 000 to over 371 000/ha depending very largely on soil, time of planting and lifting, season, variety and whether irrigation is available.

Various containers such as orchard or orange boxes and paper or woven sacks are used for despatching runners. Plants should be delivered as quickly as possible after packing. The roots of early lifted plants must be kept damp and protected, but runners lifted in the winter are generally packed direct into the containers. Excessive moisture and lack of ventilation in the package can cause heating. Delivery by direct road transport is ideal; passenger train is satisfactory in many areas, but likely times of arrival of the consignments should be ascertained and the customer notified so that rapid collection can be made. If the runners cannot be planted on arrival they should be carefully unpacked, and the bundles laid-in with soil or moist peat round the roots in a sheltered site. On no account must the plants be left in the packing containers, or allowed to get dry.

The Cold Storage of Runners

Runner plants, if lifted when dormant, can be cold stored until the following August. For fruiting the next year they must be planted in June and July when they then provide an alternative to freshly dug runners planted in August. Cold stored runners can also be used for propagation and the summer planting ensures early runner production the following year. A few plant raisers have built stores, usually of about 15 tonnes capacity, especially for the cold storage of runners. Other raisers have used small stores built for the storage of pears. Adequate insulation and refrigeration machinery capable of maintaining the required temperature of minus 1°C without fluctuation is essential.

Runners for storage must be lifted when dormant, preferably in December and January, but in some seasons lifting can continue until late February. Plants must have a period of cold weather before storing. Lifting in the second half of November is possible but the risk of deterioration in store is greater than the preferred time of December/January. Lifting in February is also possible provided the weather is cold and the plants are still dormant. Plants lifted in December and January will keep the longest.

Runners are lifted in the normal way, shaken, or washed free of soil, and 300 to 500 plants are placed in a block bottom, 38μ thickness plastic bag about 80 × 70 cm, which should be closed with an elastic band to prevent drying out. Depending on runner size, one to three plastic bags can be put in each wooden box. The boxes can then be stacked in the store, or the bags put on shelves, but care must be taken to allow a free movement of air around the bags so that the temperature of the plants is reduced to minus 1°C within a few days of storage. If the temperature of the plants drops to minus 2°C they will be damaged and a proportion will not grow satisfactorily after storage. Plant temperatures should be checked from probes placed within the bags. Store temperatures must be maintained as near to minus 1°C as possible, preferably minus 1.7°C rather than minus 0.6°C if there is any fluctuation. Infection of foliage and crown by fungi is usually the chief reason for loss of runners in store, and is often caused by slow cooling of the runners when first stored because the boxes, or bags, are stacked too tightly for correct air circulation.

Whilst plants may remain in store quite satisfactorily until early September, it is necessary to plant during June and July for fruiting (page 34) and not later than mid-August for runner production. On removal from store quick delivery is essential and the plants should be left in the sealed bags in which they have been stored and allowed to thaw out during delivery.

Tip Propagation

It is possible to root runners rapidly under a mist propagation unit in a cold greenhouse or in Dutch light frames. The mist unit is automatically controlled so that it sprays water whenever the atmosphere becomes dry and thus prevents the tiny, unrooted runner tip from drying out and dying before it has made roots. The unrooted runners with two leaves are cut off from the stolon with a sharp knife (leaving a small piece of stolon to handle the plant) and are closely placed in a suitable rooting compost, or, more conveniently, direct into a peat pot or soil block. Rooting normally takes about two weeks and then the plants need to be grown without mist, but with sufficient watering, to harden them off before planting out. Plants rooted in peat may be transplanted at a wider spacing, before planting out permanently.

This method of propagation is useful where stock is limited, as with new varieties, for it enables many more plants to be obtained from one mother plant than allowing the runners to root while attached to the parent. It may also be a useful method to provide healthy plants in disease-free compost in areas where soil diseases prevent propagation in the field.

It is important to ensure that the plants are sufficiently rooted and able to withstand field conditions before planting out or they may be severely damaged. Some planting machines will handle soil blocks or peat pots satisfactorily and the plants grow away well without a transplanting check. When planting by hand it is essential to ensure that the block or pot is in good contact with the soil, and that irrigation is available. Mist-propagated plants, because of the extra handling involved, are usually more expensive than field dug runners and are generally used by gardeners rather than commercial growers.

Micro-Propagation

When large numbers of plants are required quickly from a limited stock, as with a new variety or newly selected disease-free strain, the technique of micro-propagation in aseptic tissue culture may be the most economic and acceptable method. It is now commercially feasible to produce several million plants (it is claimed up to 25 million) from a single mother plant in one year. Thus the area of land required for stock plants is very small and production continues independently of season if suitable growing rooms are provided to bring the plants to the field stage.

A very small shoot meristem, less than 0.5 mm long, is dissected under a microscope from a stolon and grown on a sterile nutrient medium. This small meristem is either virus-free initially, or any virus will become inactivated during the culture procedure so that all the new plants will be a healthy clone. The shoot is cultured on a simple nutrient medium containing mineral salts, sugar, vitamins, auxin and cytokinin, in a screw-capped bottle placed under fluorescent lights at 25°C. In these conditions axillary buds form around the base of the shoot and these buds in turn produce more buds and shoots so that within two months the tip becomes a proliferating mass of small shoots. The shoots can be teased apart and planted on fresh medium of the same composition when further rapid proliferation will occur. When rooted plants are required, the shoots are planted on a medium not containing cytokinin and within six weeks the plants can be removed from the sterile jars and potted into compost. At this stage the humidity must be decreased gradually either by using mist or by covering with polythene, until the plants are sufficiently hardened off and have grown enough to be treated as normal runners.

The tissue cultures can be maintained on a medium which allows only slow growth without proliferation until more plants are needed when a transfer to a medium containing cytokinin will induce rapid proliferation again. Tissue culture techniques with some plant species can cause mutation and abnormalities in the progeny but this has seldom been observed in strawberries. Plants raised by micro-propagation can be entered in a MAFF certification scheme.

Varieties

The regular provision of virus-tested plants and MAFF certification schemes, together with the use of efficient aphicides, ensure present day strawberry varieties generally remain in cultivation for many more years than in the past, when virus diseases often infected all stocks. New varieties are now raised mainly at the Scottish Horticultural Research Institute, Long Ashton Research Station and the John Innes Institute. New varieties from these Stations, from private individuals and from abroad are generally first tested at the National Fruit Trials in Kent. Promising varieties are further tested at the ADAS Experimental Horticulture Stations and on growers' holdings so that their behaviour can be observed on different soils and under different climatic conditions.

Plant Breeders' Rights

The strawberry is now eligible for rights under the Plant Breeders' Rights (Strawberries) Scheme 1966 which permits breeders of new varieties of strawberries to have exclusive rights to propagate, or authorise others to propagate, plants for sale.

Of the varieties fully described in this book, the following have been granted Plant Breeders' Rights in the UK (1979):

Domanil	Montrose	Tamella
Aromel	Gento	Tantallon

Similar schemes exist in some other European countries and although each country requires a separate application to be made it is hoped that in future there will be some form of mutual agreement whereby only one applicaton will be needed. Further details concerning Plant Breeders' Rights can be obtained from The Controller, Plant Variety Rights Office, White House Lane, Huntingdon Road, Cambridge CB3 0LF.

Descriptions of recently introduced varieties, after testing at the National Fruit Trials, are published in the Leaflet HSG24 *Promising Varieties of Soft Fruits* which is brought up to date and reissued as necessary.

Choice of Varieties

The most important variety at present is Cambridge Favourite, probably comprising over 70 per cent of the total area. This variety gives heavy, regular and reliable crops and is grown both for the fresh and processing markets. The berry size does not become too small during the later picks, particularly if the plants are grown on soil well supplied with moisture. Redgauntlet, the next most important variety, is grown for dessert, particularly in Kent and Hampshire, and in areas where red core disease is troublesome. Although slightly later to ripen than Cambridge Favourite it generally produces plenty of very large fruits which aid picking and are attractive for the dessert market. Both these varieties will stand some delay in picking and remain moderately firm, even if fully ripe. They comprise the bulk of the strawberry area; other varieties being grown to a very limited extent.

Cambridge Vigour is an early variety giving at least two, sometimes four picks before the two main varieties and is popular in early areas, particularly the south-west and in Essex, but not in Hampshire where it now succumbs to red core disease. The fruit is well flavoured but rather soft and the later berries become small during the season. This variety is often grown only for annual crops. If retained for the second and third years the berries are often numerous and very small: the plants may grow too vigorously on good soils.

Gorella is the other variety that is ripe a few days before Cambridge Favourite and is now being more widely grown. The fruit is red fleshed, large, glossy and attractive but the yield, particularly in dry summers without irrigation, can be disappointing, the plants producing only few trusses, each with few berries.

All these four varieties can be grown under protection for early cropping. Redgauntlet is also used for autumn cropping as it will produce a second crop if the first crop is advanced by protection in the spring.

There is a need for a late variety to extend the season. A new Belgian Variety, Domanil, ripens after Cambridge Favourite and is being tested. Other varieties are grown only in very small quantities, mainly because they have succeeded well under particular local conditions or to meet a local demand. Amongst these could be mentioned the early variety Cambridge Prizewinner, Talisman and Tamella—two main crop varieties, and Royal Sovereign, an old variety now only grown in private gardens. In the certification scheme in 1979 nineteen varieties were listed but only four in any quantity.

Reliable autumn fruiting, or perpetual, varieties are needed particularly for pick-your-own sales; at present four varieties, Aromel, Gento, Ostara and Rabunda are available. Aromel is the only one available in the certification schemes at present. In 1977 four new varieties were selected and named after trial at the National Fruit Trials and Experimental Horticulture Stations. These have been granted Plant Variety Rights and may become commercially available over the next few years. One of them, bred at the John Innes Institute, has a potential for mechanical harvesting and is named Harvester. Three varieties raised at the Scottish Horticultural Research Institute, Tantallon, Troubadour and Saladin, all have field resistance to red core disease.

Description of Varieties

Major Varieties

CAMBRIDGE FAVOURITE (*Also suitable for processing, protection and pick-your-own*)

The most widely grown variety because of its heavy and reliable cropping capacity. The fruit remains firm for several days and does not soften or become dark in colour, valuable assets if picking is difficult. It travels well. Not susceptible to mildew, but susceptible to *Botrytis* in wet seasons. Not so susceptible to verticillium wilt as some other varieties, except in East Anglia. Not field resistant to many strains of red core. Susceptible to red spider mite. Tolerant of virus diseases carried by aphids but isolated plants frequently infected with green petal virus. Streaky yellows very occasionally seen. Produces runners profusely (Plate I, top). Makes a dense plant on all soils, but the fruit is easily seen and picked.

Raised by Mr D Boyes at Cambridge
Plant fairly large, moderately open, vigorous, spreading.
Flowers large, mainly protected by the foliage.
Fruit large, maintaining size well over the picking season, rounded, plump; pinkish-scarlet, turning light red, colouring evenly, somewhat dull. Flesh firm, moderately juicy, pale pink to pale red, moderate flavour. Plugs fairly easily.
Season second-early to mid-late.

REDGAUNTLET (*Also suitable for processing, protection and pick-your-own*)

A moderate to heavy cropper of good habit that maintains good berry size. Plant growth is sometimes weak, particularly under poor soil conditions. The fruit is well exposed, making picking easy but the fruit stalk is sometimes tough. Travels well and is of reasonable market appearance. Field resistant to some strains of red core and to verticillium wilt and not susceptible to mildew or *Botrytis*. This variety has the ability to initiate trusses during the summer, particularly if the first crop is protected in the spring, and then flower in August and will crop satisfactorily, generally with some protection, from mid-September until November, depending on the season. Occasionally in very warm years this second crop can be picked in the autumn from unprotected plants. Flower initiation and second cropping is more marked in the early areas. This second crop does not appear to reduce the crop the following year from normal, autumn initiated trusses. Less prolific in runner production than Favourite. This variety responds well to warm conditions and sheltered sites.

Raised in 1946 by Mr R D Reid, Scottish Horticultural Research Institute; introduced by the Institute in 1956–57.
Plant vigorous, tall and spreading with sparse foliage, rather slow growing in the spring.
Flowers medium size, only partly protected by the foliage.
Fruit large to very large, rounded to spherical-wedged, broad, blunt, often uneven, with few seeds, attractive scarlet becoming dark crimson. Flesh firm, fairly juicy, deep scarlet, only fair flavour. Plugs fairly easily.
Season mid to mid-late, but early under protection.

CAMBRIDGE VIGOUR (*Also suitable for protection and pick-your-own*)

A fairly heavy cropper of good quality from one-year plants. Field resistant to some strains of red core, but very susceptible to verticillium wilt and mildew. Sensitive to drought. Attractive fruit that does not travel well and is inclined to be soft if at all over-ripe, so needs more frequent picking than Favourite. Older plants produce far too many small berries. Produces plenty of runners and may grow too vigorously on good soils.

Raised by Mr D Boyes at Cambridge
Plant large to very large, upright and spreading, very dense.

Flowers numerous, medium-large, well protected by the foliage.
Fruit fairly large but inclined to be very small towards the end of the picking season, sharply conical, regular, sometimes a little wedge-shaped, very attractive orange-red to light red, turning scarlet. Flesh soft, juicy, moderately sweet, good flavour. Plugs fairly easily.
Season fruit from maidens ripens early, but fruit from older plants is mid-season.

GORELLA (*Also suitable for protection and pick-your-own*)

This variety produces large, early, glossy fruit. Yield is not as high as the three varieties already described and irrigation is often required. The habit is compact, and the foliage sparse and susceptible to powdery mildew. There is field resistance to some red core races but not to verticillium wilt. Botrytis fruit rot is generally not a problem. Under protection the primary fruits can be too large for easy grading and packing.

Raised by Miss H Kronenberg at the Institute for Horticultural Plant Breeding (IVT), Wageningen, Netherlands, and introduced in 1960.
Plant moderate growth, sparse, with coarse thick foliage. Leaflets often five instead of three per leaf.
Flowers mid-season, medium size, on thick stalks.
Fruit very large, conical, crimson-red, sometimes with a green tip, well exposed beneath the upright foliage. Flesh bright crimson, juicy, average flavour.
Season second early.

Varieties of Minor Importance

CAMBRIDGE PRIZEWINNER (*Also suitable for protection and pick-your-own*)

Essentially an early variety with very attractive fruit but not a heavy cropper. Susceptible to verticillium wilt and red core, but berries not very susceptible to botrytis fruit rot. Rather shy in producing runners. Needs good soil to grow and crop satisfactorily.

Raised by Mr D Boyes at Cambridge
Plant medium size. Growth moderate to weak, according to locality where grown, and compact.
Flowers medium size, somewhat exposed.
Fruit conical, occasionally wedge-shaped, bright red, good colour, large at first but rapidly loses size. Flesh only moderately firm; pale red, sub-acid of moderate flavour, plugs easily. Calyx large.
Season early.

TALISMAN (*Mainly pick-your-own*)

At one time widely grown, both for dessert and processing markets but the lack of berry size is now the chief disadvantage. Field resistant to red core and mildew. Runners profusely. Grown for late ripening fruit for pick-your-own.

Raised in 1946 by Mr R D Reid, Scottish Horticultural Research Institute; introduced in 1955.
Plant medium size, very slow to develop in the spring, upright, dense, fairly vigorous.

Flowers smallish, usually fairly well protected by the foliage.
Fruit conical to wedge-shaped, medium sized at first, becoming smaller as the season advances. Fairly attractive bright scarlet turning dark scarlet when fully ripe. Flesh firm, juicy, scarlet, fine texture, good flavour. Plugging fair.
Season mid to late.

DOMANIL (*Also for pick-your-own*)

A late, heavy cropping variety which may bring a useful extension to the fresh fruit season. There have been cases of genetic breakdown in this variety resulting in June yellows; it is hoped that an improved clone will be produced by re-selection. Susceptible to mildew.

Raised in 1960 at Station Fruitière et Maraichère, Grand-Manil (Gembloux), Belgium.
Plant very vigorous, tall, fairly dense, and of a spreading habit. The leaves are medium to dark green, rather crinkled and very glossy.
Flowers medium size, flattish, and well hidden beneath the foliage.
Fruit large size, round conical, and orange-red in colour, moderately well displayed. The flesh is orange, firm and of moderate flavour. Fruit easy to plug but because of its large size more suited to the fresh market than for canning or freezing.
Season late.

MONTROSE (*Also suitable for protection*)

A very high yielding variety ripe just after Cambridge Favourite. It is not resistant to red core or powdery mildew and runner production can be poor.

Raised in 1960 at the Scottish Horticultural Research Institute.
Plant similar in appearance to Redgauntlet but more vigorous. Tall and medium dense.
Flowers larger but fewer than with Redgauntlet.
Fruit medium to large in size, the berries are well displayed. Paler than Redgauntlet and more rounded. Moderate flavour, juicy and fairly easy to plug, but rather soft.
Season mid.

TAMELLA

A heavy cropping, second early variety which has the drawback of being susceptible to *Phytophthora cactorum* (causing rotting of the crown).

Raised in 1962 at the Institute for Horticultural Plant Breeding (IVT), Wageningen, Netherlands.
Plant medium vigour and height, and moderately dense.
Flowers medium size, cupped, and held level with the leaf canopy.
Fruit large size, long conical in shape, and medium red colour well displayed. Flesh is a medium orange-red, very juicy, of medium firmness, and moderate to good flavour. Rather difficult to plug and the core is sometimes removed at the same time. Rather soft and only really suitable for the fresh market.
Season mid.

ROYAL SOVEREIGN

A very old variety introduced 1892, and very susceptible to virus so that planting and maintenance of healthy stock is essential. Generally considered one of the best flavoured varieties but only gives light crops.

Fruit bright, pinkish-orange in colour. Very susceptible to red core, *Botrytis*, mildew and virus diseases. Now only of interest to gardeners.

TANTALLON

Raised at Scottish Horticultural Research Institute; introduced 1978.
Plant medium vigour, spreading rather flat, light green.
Flowers medium, fairly well protected by the foliage.
Fruit medium, well displayed, round to round conical, orange-red with red flesh, sometimes rather soft, flavour moderate.
Season second-early to mid-late very similar to Cambridge Favourite. This new variety has field resistance to red core but is susceptible to *Verticillium* wilt. It is suitable for tunnels. It does not produce any second crop. In trials has out-yielded Cambridge Favourite. Being entered in MAFF Certification Scheme.

Perpetual Varieties

At present certified stock is not available for these varieties though it is hoped virus-tested plants will ultimately become available so that some perpetual varieties can be included in the Scheme. All perpetual varieties are shy in producing runners which makes them more difficult to propagate than main season varieties. It is important to obtain runners as healthy as possible. Under certain conditions Redgauntlet can be made to produce a second or autumn crop by forwarding the first crop (pages 7 and 70).

AROMEL

Raised by Mr R Cumberland of Poole, Dorset.
Plant moderate to vigorous and of a habit more upright than spreading.
Flowers well displayed.
Fruit medium sized, red in colour with medium red flesh, well displayed. Rather soft but of very good flavour.
Runner production is only moderate but better than either Ostara or Rabunda.

GENTO

Raised by Hummel Ltd., Stuttgart, Germany.
Plant moderately vigorous, upright spreading.
Flowers very well displayed on thick stalks.
Fruit medium to small sized, blunt conical, reddish-crimson, red flesh, moderately firm and of good flavour. The calyx parts readily from the fruit. This variety can produce large berries for an autumn fruiting variety but cropping is variable. Gento is less perpetual than the other three varieties listed here tending to produce a second or autumn crop rather than cropping continuously.

OSTARA

Raised at the Institute for Horticultural Plant Breeding (IVT), Wageningen.
Plant of moderate vigour and height, with a rather open and spreading habit.
Flowers small to medium in size, and saucer-shaped.
Fruit small to medium in size, conical and red, well displayed. The flesh is orange, moderately firm and of a moderate flavour. It is very easy to plug.

RABUNDA

Raised at the Institute for Horticultural Plant Breeding (IVT), Wageningen.
Plant of moderate vigour and height, rather dense, and of a habit intermediate between spreading and upright.
Flowers of medium size and flattish.
Fruit medium to large in size, conical, and orange-red in colour. The flesh is pale orange, rather soft, with a moderate to rather poor flavour. The berries are rather hidden in the first flush but more visible in later flushes. Plugging is fairly easy.

Varieties for Processing

Most varieties are suitable for jam making and supply and price are more important factors than choice of variety. Higher prices may be offered by processors to growers contracting all or part of their crop for specific varieties for quick freezing or canning. Processors may, in addition to specifying variety, also lay down conditions relating to the spray chemicals that can be applied, to avoid taint problems. Contract growing for processing is sometimes not so profitable as growing for the dessert market; however, if high yields can be obtained, contract strawberry growing for local factories, which avoid all marketing problems, is well worth consideration. Many growers offer their early picks of larger berries on the open market and the main picks to the factories. The standard of fruit required for quick freezing, canning and bottling is much higher than that needed for jam or pulp, but jam contracts may now specify dessert quality fruit.

Canning, Bottling and Quick Freezing

For canning or bottling strawberries need to be of a medium, uniform size, with a good flavour after heating. It is an advantage if the core remains in the fruit after plugging as this ensures a firm texture and assists in retaining the shape of the berry after processing. The core (plug) should be barely distinguishable from the flesh and never tough. The berries should also take up the artificial colouring normally used to give an attractive product.

Though no variety now grown can be considered perfect in all respects for canning, Cambridge Favourite is one of the best, giving a well-coloured, pleasantly flavoured product. Cambridge Vigour is sometimes inclined to be dull in colour but otherwise satisfactory and, like Cambridge Favourite, the berries are easily plugged. Redgauntlet has proved fairly good though the exceptionally large size of some of the berries from spaced plants and the tendency to break down during processing can be disadvantages. However, from matted rows Redgauntlet will produce a more suitable berry.

For quick or deep freezing, fruit is needed which will retain a good colour and flavour after thawing. Cambridge Vigour and Cambridge Favourite are both acceptable although the latter, which is the most widely used variety commercially, is really too pale for this purpose.

Climate, Site and Soil

Climate

The strawberry, like other fruit crops, is susceptible to spring frosts. The flowers open in sequence on the trusses and the plant has a long flowering period during April and early May, so that a complete kill of all flowers due to frost damage is unlikely. If the temperature falls low enough for long enough, the centre of the open flowers containing the incipient fruit is damaged and shows the characteristic 'black-eye' and the crop may be seriously reduced. Unopened flowers and those that have set are slightly more resistant to frost than partially and fully opened flowers, but can be damaged if the temperature falls low enough.

It is important therefore to select sites not subject to spring frosts and particularly to avoid valleys and frost pockets. On still, clear nights radiation frosts can occur; the soil and the plants lose heat by radiation into space and the air in contact with them is cooled. This process may continue until the air temperature falls below freezing. If the land is sloping the dense, cold air moves slowly downhill collecting in valleys (frost hollows). Frost pockets may be caused by windbreaks such as woods, hedges or walls which prevent the slow flow of colder air away to lower ground. Thus plants at the bottom of the field may be frosted whilst those higher up the slope may not. Low areas of ground will collect the dense, cold air while fields on higher or sloping ground may escape.

Freezing can also occur during windy conditions when the temperature is low and these wind frosts are much more damaging to exposed fields than to those that are sheltered. However during April and May radiation frosts are more likely to occur than wind frosts.

Large areas of water, as sea or as reservoirs, may create locally favourable situations. Many of the early strawberry areas are situated in sheltered sites within a few miles of the sea, where the spring sunshine is brighter, and warmer than inland areas.

Site

Apart from the risk of spring frost the strawberry always grows and crops best under sheltered conditions. Wind can seriously reduce growth. A long growth period with an early spring, warm summer and mild autumn is desirable and provided there is no risk of spring frosts, sheltered sites with gentle south-westerly slopes are ideal. A site sheltered from the wind but exposed to the north may be satisfactorily used to obtain worthwhile crops from a late variety.

The growing season in the south-west and south of England is longer than in the east and north, but within any major area there can be marked growth variation depending on slope and aspect from field to field.

Shelter belts can be planted to provide wind protection to otherwise favourable sites. Retention and encouragement of existing hedges and woods is often worthwhile.

Strawberries are sometimes grown interplanted between other fruit crops. Planting between cane or bush fruit, or intensive systems of apples and pears is unwise and not recommended. Intercrops of strawberries under wider plantings of top fruit are more acceptable but difficulties from spraying and other operations are bound to occur and it must be remembered that the permanent crop should not be neglected in favour of the immediate needs of the strawberry crop. Whenever possible interplanting should be avoided.

Soil

The strawberry will succeed on a surprisingly wide range of soils, from heavy clays to light sands and gravels. On very light soils irrigation will be necessary to establish runners when planted early and to maintain berry size and quality during the picking season. On most loams and clay soils, provided the plants have rooted well and there are no root disease problems, summer moisture reserves are likely to be adequate, except in exceptionally dry seasons such as 1975 and 1976. But for maximum maincrop yields the heavier soils with good reserves of moisture are better than lighter soils. The silty loams and sandy clay loams of East Anglia can carry high plant populations without irrigation and in this area matted rows are grown to obtain high yields of medium sized fruit for processing.

Light soils have slightly higher temperatures in the spring than heavier soils and growth will start a little earlier; they may also produce a smaller plant which often flowers earlier. These may be valuable assets and, provided irrigation is available, good yields can be obtained. Expenditure on bulky organic manure to improve lighter soils for strawberry production may be well worthwhile.

In some areas very stony soils are popular for strawberry growing; the stones act as a mulch to retain moisture, yet heat rapidly to warm the adjacent air. They also provide such a stony surface that strawing is unnecessary. Herbicides can now reduce the cultivation difficulties previously associated with very stony soils.

The strawberry will not tolerate impeded drainage or waterlogging and plants will be killed. Well-drained fields are required and it is important to keep ditches cleared and outlets from tile drains should be examined each autumn and cleared as necessary. Compaction of otherwise well drained soil through misuse of machinery under wet conditions before planting may cause waterlogging or a soil pan and root damage. In some areas where there is a high water-table during the winter, this may cause death of the deeper roots. Even if plants are not killed by waterlogged conditions, a part of the root system may be damaged, resulting in poor growth and crop and a greater susceptibility to red core disease.

Soil infected with verticillium wilt or red core disease (see Pests and Diseases, page 84) which reduces the normal rooting, will give a poor crop or complete failure. Certain soil-living eelworms can transmit virus diseases and their presence, if suspected, should be checked by tests before planting. Strawberries are often chosen as the crop to follow cleared woodlands and there is a particular danger from eelworms under these conditions. If no eelworms are present, cleared woodland soils rich in humus are very suitable for strawberries.

On alkaline soils and particularly those derived from chalk, or containing much free chalk, strawberry plant foliage may become chlorotic, due to lack of iron absorption by the roots. Provided there are sufficient moisture reserves and good depth of rooting, moderate chlorosis does not affect cropping.

Strawberries can follow many crops in the rotation though potatoes are not suitable as there may be verticillium wilt left in the soil. Winter harvested crops such as sugar beet and winter vegetables often leave the soil in poor condition. Corn crops and leys are suitable. On the larger holding it is well worth planning a rotation to leave the soil as weed free as possible. It is **strongly recommended** that the rotation be planned to allow a pre-planting herbicide fallow. This will permit the use of herbicides (page 46), approved for the control of perennial and annual weeds on bare soil, which it would not be possible to use once the strawberries are planted. This pre-planting cleaning period is particularly recommended if strawberries are to follow permanent pasture, over-wintered arable crops, or cleared top or bush fruit. Sub-soiling and correction of any deficiencies of nutrients in the lower, as well as the top soil can also be done and the land left in an ideal condition for planting. The local ADAS Horticultural Adviser should be consulted and can arrange for soil testing for nutrients, or specific problems.

Manuring

As the strawberry may be cropped for one year only, or for up to three years or more, manuring should vary according to the time the plantation is to be retained. Before planting, a sample of the soil from the site should be analysed and an estimate made of the requirement of lime, potash, phosphate and magnesium. If any soil deficiency of potassium, phosphorus or magnesium is remedied prior to planting, further fertiliser applications to one-year-old plants will not be needed. A pre-planting dressing of farmyard manure (FYM) often provides all the nutrient needs of the crop over several years on fertile soils. The availability of nutrients is influenced by root development which itself is dependent on the depth, texture and structure of the soil and the availability of water.

Nitrogen (N)

While nitrogen is necessary for plant growth, the strawberry is able to obtain most of its needs from the release of nitrogen during the breakdown of organic matter in the soil. Experiments using virus-tested clones have not shown much response in cropping to nitrogen fertilisers in their early years. Too much nitrogen may be detrimental to strawberries: it can give excessive vigour and this may lead to a reduction in crop because fewer flower trusses are initiated. There is no experimental evidence that organic forms of nitrogen such as hoof-and-horn or meat-and-bone manure have any advantage in place of inorganic forms.

Phosphate (P_2O_5)

Experiments have shown that on most soils where strawberries are likely to be planted in this country the phosphate requirement of strawberries is small except in the year of planting.

Potash (K_2O)

The importance of potassium for strawberries has been shown in many experiments. On very potassium-deficient soils, where farmyard manure has not been applied, the omission of potash has generally had serious effects on vigour, longevity and cropping. Annual applications of potash fertiliser are therefore needed, especially when adequate dressings of farmyard manure have not been used.

ʼlagnesium (Mg)

he magnesium requirement of strawberries is small, but a deficiency sometimes ccurs on sandy or light soils overlying gravels, or where excessive applications of otash fertilisers have caused an imbalance between the levels of magnesium and otash.

ime

he strawberry is tolerant of fairly acid soil conditions, but the pH should not be lowed to fall below 6.0. Plants can fail or be much less vigorous on a calcareous oil due to deficiencies of iron or occasionally manganese. Manganese deficiency ccurs especially on soils with a high organic matter content and where the soil H is 6.5 or more. Liming should therefore be used carefully and be restricted to cid soils; ADAS advice should be sought if in any doubt and the soil sampled nd analysed before lime is applied. It is always preferable to correct any lime eficiency by liming the crop preceding the strawberries.

lanuring Programmes

efore Planting

dressing of bulky organic manure is beneficial. FYM is suitable at rates of 50 to) t/ha. Every tonne supplies approximately 1.5 kg N, 2 kg P_2O_5, 4 kg K_2O and 8 kg Mg. Deep litter or broiler house poultry manure can be used but as they ontain more nitrogen, phosphate and potassium than FYM, they must be used at wer rates. Two to three tonnes of deep litter or broiler house manure are uivalent to 20 tonnes of FYM. If annual cropping is practised on the same site application of bulky manure will be needed every three to four years.

Any mineral deficiency shown by soil analysis should be corrected by applying e appropriate amount of fertilisers, as shown in Table 1 (after allowing for itrients derived from any FYM application). If the pre-planting soil nutrient dex is 'O' it is wise to plough in half the fertiliser and cultivate the other half into e top soil before planting.

fter Planting

o further fertiliser will be required for annual cropping, but for plantations to be tained the necessary amounts should be applied each year after picking and eaning up the beds (Table 1). The plants are initiating flower trusses for the next ar's crop at this time and adequate nutrients are essential; spring application of rtilisers is not recommended.

Nitrogen should only be applied to healthy strawberry beds with caution as cess nitrogen may encourage leafy plants and excessive runner production. On any good deep soils there will be sufficient natural nitrogen available, articularly if bulky manures have been used pre-planting. If plants are retained r three crops or more, nitrogen at 40 kg/ha may be required annually after

---

F1, *Fertiliser Recommendations for Agricultural and Horticultural Crops* or GF23, *Lime and rtiliser Requirements* No 3 *Fruit and Hops*, give further details of soil sampling and manuring.

harvest: this will depend on the growth of the plants, especially in low rainfall areas (rainfall less than 350 mm per year).

Table 1

Amounts of plant nutrients to be applied (kg/ha) according to soil analysis

P, K or Mg Index	P_2O_5	K_2O (kg/ha)	Mg
0	110	220	60
1	70	150	40
2	**40**	**80**	30
3	40	Nil	**Nil**
Over 3	Nil	Nil	Nil

In the absence of results of soil analyses use the figures in bold print.

Runner beds

It is very important that runners should be produced in soil which is well supplied with nitrogen, phosphate and potash. The soil should also be supplied with organic matter to help in maintaining soil conditions suitable for rooting and satisfactory growth of runners. Unless the soil has a high organic matter status, is a peaty soil or irrigation is available, FYM can be highly beneficial to runner production and should be ploughed in before the final preparations are made for planting in the autumn. Where adequate irrigation is available the provision of organic material is of less importance.

It is very important that the developing runner plants have a good supply of plant nutrients particularly if they are to be used to crop the first year after planting, since the flower trusses are formed at an early stage within the runner. Good growth after planting in the permanent position will not compensate for starvation in the runner bed.

The soil should be analysed prior to planting the mother plants and the appropriate amount of potash, phosphate and magnesium applied depending on the soil nutrient index before planting (Table 1). Nitrogen, which is readily leached from the soil, can be applied in the spring to a runner bed, if bulky manures have not been used.

Planting Systems

Spaced Plants

Strawberries are traditionally grown in the field either as spaced plants or in matted rows. In the spaced plant method all runners are removed during the life of the plantation; the plants are usually spaced 0.3–0.5 m apart in rows 0.9–1 m apart. These distances allow pickers to work between the rows and the plants on good soil generally meet along the row. On really good soils it may be wiser to space the rows wider than 0.9 m.

Spaced plants intended for one crop only are usually planted closer, generally 0.3 m apart in the rows, but unless fairly weak growth is expected it is not wise to reduce the distance between the rows much below 0.9 m or pickers will damage the fruit. The chief advantages of the spaced plant within the row is the production of bigger individual berries and the ease with which fruit is seen and picked. The chief disadvantage is the cost of runner control. There is little to be gained by planting 0.3 m apart in the rows for the first crop and spacing out plants to 0.6 m for subsequent crops.

With the increasing use of efficient herbicides, and hence less need for cultivation, row and plant patterns can be varied. On good soil these distances do not always leave adequate room for pickers and some fruit damage results; yet wider spacing tends to reduce yields, particularly in the first year. Growers should plan their planting distances to fit, not only the soil and variety grown, but their machinery and particularly the method to be adopted for picking. The use of a picking tray on legs over the row (page 76) may reduce damage during picking. Another possibility is to vary the distance between rows, allowing a narrower alley between every other row and picking the two rows from the wider alley.

Matted Rows and Beds

In the matted row system all runners are allowed to root along the row, but growers try to limit the number of plants by light cross harrowing over the rows and by some removal of excess runners in the alleys. The original rows are planted about 0.9 m apart, with 0.45 m between the plants. As the runners develop during July and August in the first year they are brushed into the row, usually by women using short sticks. These are allowed to root, making a mat of runners about 0.4 to 0.45 m wide. Any extra runners are cut off in the autumn by rotary cultivators, or destroyed by directed sprays of paraquat. Flailing of the beds with a forage harvester, after harvest but before the application of residual herbicides, followed by the removal of debris, also clears away excess runners before they have time to root.

In the matted rows, as there are more plants per area and a proportion of these each year are one-year-old plants, the system can, under suitable conditions, give a higher crop over a longer picking period than spaced plants. This system is very suitable for obtaining heavy crops of medium sized fruit for processing. However, overcrowding of plants in matted rows results in severe competition for water in dry seasons, and this system of growing is not suitable on shallow soils, or those without good soil moisture capacities. Overcrowding may also result in an increase in fruit rots and consequently reduced yields.

With the advent of herbicides there has been some interest in matted beds. In this system two rows of plants are allowed to run into each other and completely fill the inter-row area to make a bed 1.2 m wide. A path of 0.45 m is allowed between the matted beds. High yields can be obtained by this method but very careful attention to manuring and disease control is necessary and picking across the bed is difficult.

Both matted rows and matted beds are often given a routine mowing or flailing of the foliage after harvesting. The debris is then either raked, blown or burnt off, or maybe incorporated in the soil by inter-row rotary cultivations which also remove unwanted runners.

Large numbers of runners are not normally produced after the first year, but both matted rows and matted beds can become too crowded. Growers should check the number of runners being produced and with this information and experience may need to vary the planting distances for matted rows to fit the expected growth of each variety on each soil type, and the spray machinery being used.

It is important partially to control runners, particularly the second year's growth, in matted rows, or the crowns will become too crowded and poorly rooted. Runner production is less prolific from plants that have cropped heavily.

For the tractor mounted, mechanical harvester a matted bed to fit the cutter bar width will probably be the best system, so that the harvester can cut off and elevate the maximum number of fruiting trusses at each pass. Row widths will have to fit the wheels.

Row Lengths and Position

A recent work study indicated that strawberry rows should not be longer than about 65 m. This reduces to a minimum the time pickers spend walking with their fruit to the tally or collection point. Much fruit is picked by piecework and as pickers are generally in short supply they should be helped to spend the maximum amount of their time in picking. Long fields should be divided by cross alleys every 130 m, allowing picking from each end. Collection trailers can be stationed in the cross alleys. The position of moveable shelters within the field, both for pickers and fruit stacking and shading is also worth study. Planting is done only once and, with the advent of herbicides, cultivations are less laborious. It may therefore be better to plan the rows to help harvesting operations, even if this means shorter tractor runs for other operations. Row lengths should be very carefully planned for pick-your-own fields, as long rows are totally unsuitable and result in much fruit being left unpicked on distant plants. Plenty of cross-paths, to encourage pickers to walk to suitable areas, should be provided from the car parks and headlands.

Drainage is important and the rows should not restrict any natural drainage; if plants are on ridges these will have to run with the slope of the field. On slopes where erosion may be a problem the plants can be planted across the slope and will help to stabilise the soil provided the tractor can work that way.

Number of runners to plant area

Planting Distance m	Plants ha
0.9 × 0.3	37 000
1 × 0.5	20 000
1 × 0.6	16 700
1 × 1	10 000

Time of Planting

Runners can be planted at any time from July until the following April or May, provided soil and weather conditions are suitable. If a good first year crop is required early planting is essential particularly so in less favoured areas. Early plantings may require irrigation, as runners lifted from July to September are actively growing and will not withstand strong sunshine or dry soil conditions. Runners established before the beginning of September will yield well the first year in most areas; those planted during September may give moderate crops, very much depending on the autumn weather after planting, and the locality. Provided there is no check to establishment, the size of runner does not matter for early planting since the soil is warm and root growth rapid, but larger runners will grow and crop better than small runners if planting is not possible until mid-September or later.

Runners planted in October and November seldom need irrigation and will readily become established. They grow better than those planted in spring, but generally will not make plants strong enough to give a worthwhile crop the first summer. Mid-winter planting is not recommended; the runners are best left in the runner bed and planted as soon as possible in the spring.

Runners planted in the spring and deblossomed will make strong plants for first cropping as two-year-old plants. After cold winters runners may be found to have been lifted by the frost and need re-firming in the spring, either by foot, or using a lighter roller.

Some growers planting in late winter or spring space plants 0.6–0.9 m apart in the row. These plants should be deblossomed the first year, and an adequate plant population is obtained by allowing the new runners to root in the rows. But wide plantings demand few establishment failures or a gappy field will result; the advantages are cheaper establishment costs, both in the purchase of runners and in planting them.

Annual Cropping

Growers who raise their own runners have a big advantage in that they can time the lifting and transplanting to fit the weather conditions. Most growers of annual crops of strawberries have to produce their own runners in order to ensure August planting and worthwhile yields the following year. It is most important that growers raising their own runners should use certified mother plants and take every care to avoid pest and disease problems during propagation and carry out all the precautions described in the chapter on 'Propagation and Need for Healthy Plants'.

The strawberry flowers earlier and the fruit ripens a few days earlier on one-year-old plants than on older plants, the berries are usually larger, of good quality and easily seen and the plants can be readily sprayed and strawed. On sheltered sites, using early ripening varieties such as Cambridge Vigour, Cambridge Prizewinner, Gorella or even the standard variety Cambridge Favourite, one to three picks will be obtained from one-year-old plants before fruit on older plants is ripe. The price premium obtainable for this earlier fruit, which fills the gap between the end of the protected crop and the main field strawberries, may make it worthwhile to plant every year for this early ripening, good quality fruit.

Annual cropping reduces problems from pests and diseases, and there is no expense in runner control. However there are heavy establishment costs in producing, or purchasing, early runners and in replanting each year.

In some early areas where red core and/or verticillium wilt are endemic and where strawberries have been grown with little or no rotation for many years, annual replanting is essential since plants do not grow well after the first crop. Planting a strong, well-rooted runner from clean land, during August, or early September, enables the plant to survive these unsatisfactory soil conditions, to make moderate growth and produce sufficient early fruit to warrant the high growing costs. In these areas most of these plants are protected for earlier crops with consequential higher prices (see chapter on 'Early Strawberries').

Use of Cold Stored Runners

The chief value of cold stored runners is that they provide a guaranteed supply of certified runners for the early summer planting for annual cropping (see preceding section) and for early production (page 59). Whilst freshly dug runners are often planted for this purpose in August, they then are expensive and may be in short supply. Cold stored runners need to be planted earlier than freshly dug runners to get strong growth and comparable yields in the next spring. Experiments have shown that cold stored runners should be planted before early July in the north and east of England, before late July in the southern areas and before early August in the very warm, south-west. After these dates the crop from cold stored runners is much reduced and in some cases the plants will fail to initiate any trusses and be blind the following spring. The cold stored runners produce flowers within a few weeks of planting and these must be removed, unless the plants have made very strong growth. If these runners are not deblossomed and allowed to carry a light autumn crop, the plants have then to initiate more flower trusses for the crop the next spring and yield in the spring is reduced.

Cold stored runners are occasionally planted in May and June and allowed to flower and crop. But this first summer crop is always light and generally ripens during the main production period in July. It is difficult to get good crops of ripe fruit in August by later plantings without depressing the more worthwhile crop the next spring. There is little to commend the planting of cold stored runners in May and June and deblossoming. Plants so treated will make good growth and during August and September, produce plenty of runners which can be laid in to give a well matted plantation the next spring, but the cold stored runners are more expensive than runners direct from the runner bed. It would be more economic to plant freshly dug runners in April for cropping the next year, rather than cold stored runners a few weeks later, provided the land was available.

Because cold stored runners have to be lifted and placed in store in January, definite orders must be placed then with the runner producers for delivery the following summer. The cost is usually more than that of runners delivered in the spring but less than early (August), freshly dug runners. The runners thaw out in transport from the store and they must be planted immediately on arrival and never left to sweat and heat in the polythene bags. Further, it is unwise to consider the use of cold stored runners unless irrigation is available before, and immediately after, planting. Cold stored runners, once they have thawed out, are very prone to desiccation and need moist, warm soil to make new roots rapidly. Cold stored runners are more susceptible to damage from herbicides than freshly dug runners planted in August and September. Only local experience can show if herbicides can be safely used a few days after planting, or whether a longer period should be allowed.

Certified cold stored runners also offer a certain supply of mother plants for early planting for propagation purposes and if properly managed will definitely give many more early runners the following year than open ground plants. (STL 54. Cold Storage of Strawberry Runners and their Use, Appendix 1).

Cultural Operations

Soil Preparation

Thorough soil preparation is important; it is completely uneconomic to plant strawberries on sites infested with perennial weeds such as docks, thistles, bindweed, couch grass and perennial nettles. It is impossible to eradicate these later without considerable damage to the plants. Ideally, a spring or summer fallow should precede early planting during August to September. If a ley is to be ploughed out for strawberries it should first be rotary cultivated or disc harrowed and 50 kg/ha of any type of nitrogen fertiliser applied to aid rotting, and then ploughed in not later than early June. When strawberries are to be planted in the autumn, winter or spring following an arable crop, the soil should be ploughed as soon as possible after removal of the crop in summer. If perennial weeds are present, adequate cultivations, or preferably the use of suitable herbicides in a fallow, must eliminate these before planting. Certain herbicides require a minimum interval between application and planting.

It is important not to cultivate or plough land for strawberries if conditions are not correct. Many cases of poor growth of strawberries have been shown to be due to pre-planting soil preparation under wet conditions. Strawberry roots are very intolerant of soils with poor structure, or where impervious layers of soil have been produced by ill-timed cultivations. For the same reason it is essential to ensure good drainage and to break up any plough pans. Subsoiling, preferably with a winged subsoiler tine, should be done under as dry conditions as possible, well before planting. A chisel plough is a useful tool to ensure thorough soil preparation.

As the strawberry roots are planted 15 cm down and will continue to grow downwards, it is wise to correct any obvious deficiency in soil nutrients, as shown by a soil analysis (see chapter on Manuring) by ploughing in half the recommended fertiliser dressing for phosphate and potash and applying the rest on top before planting. Neither element is readily leached by rain and the movement of both can be slow in the soil.

If farmyard or other bulky manure is being used this is best worked into the top soil and not ploughed in deeply to form a mat. FYM can be applied after ploughing and the cross-cultivations will work it into the top soil as the furrows break down or, if the conditions of the previous crop permit, it can be ploughed in with a partially inverted furrow.

After ploughing, the soil should be cultivated to produce a rough tilth and thereby ensure a good germination of the annual weed seeds. These can then be killed by light cultivations, or alternatively by a desiccant herbicide such as paraquat, used a few days before planting and without further cultivations. This method ensures the minimum of weed seedlings during the first few weeks after planting while the strawberry plants are becoming established. For a complete herbicide programme see the chapter on 'Weed Control with Herbicides and Cultivations'.

It is particularly important on clay and silt soils not to consolidate the surface soil too firmly. The roots of the newly planted strawberry need oxygen as well as soil moisture and warmth. Provided there are no problems of drying out, planting in slightly loose soil is preferable to planting in over-consolidated land, particularly if irrigation is to be applied.

It is unwise to prepare too fine a surface tilth before planting, particularly if a system of no cultivations is to be adopted. Frequent cultivations to obtain a tilth will encourage erosion problems on sloping sites and may also cause silting over the crowns of plants set too deeply. A reasonable surface with sufficient depth for working the planter, or to plant by hand, is all that is required.

In some areas where drainage is poor, or red core endemic, strawberries are grown on ridges. The usual practice is to make the ridges with potato ridgers, roll lightly and then plant by hand or machine on the ridges. The ridge should be not too high or pointed, but a rounded mound.

Marking Out

If a machine planter is used it is only necessary for the driver to mark a sight line across the first row—or line—with poles in the usual manner. The planter has a setting arm that will mark out for subsequent tractor runs across the site.

If planting by hand on a small scale, pegs and lines of string may be used. On a larger scale, one line can be so laid out and a marker used to draw out further lines across the field. The marker has three tines set in the cross-piece to mark three row distances and the end marker runs in the last row marked. A tractor can similarly be used to mark larger areas, setting pointed tines the required distance apart on the cultivator bar and sighting the first run with poles. The tractor wheels should be set to avoid running on the planting lines and so over-firming the soil.

It is unusual now to plant on the square, as was once done to allow cultivations in both directions. The distance between plants can be gauged by eye, if planting by hand. With some planting machines the setting distance can be automatic and is adjusted to the tractor speed.

Planting

Whether planting by hand or machine, it is important to prevent the roots of the runners drying out. This is especially necessary if planting during July to early September, when the runners are actively transpiring, and almost as important in the spring when searing winds can desiccate the fine rootlets. Thus if a gang is planting and some workers are dropping plants ahead of the planters at the correct spacing, they should not be allowed to get too far ahead. Runners should be kept covered with damp sacks, or in their original packing material if suitable. A good damping over-night is sensible and they should be protected from wind at all times. Out in the field runners should be kept protected in trays or sacks and on the planting machine only sufficient runners for one journey should be carried.

The larger the runner and the bigger the leaf surface, the more it will lose water through transpiration. This suggests that cold stored runners with little or no leaf are not so sensitive to planting conditions—until they start to grow.

Planting should be done only under good conditions. In late summer and early autumn it is unlikely to be too wet as soil moisture reserves will not have been replenished; the soil is then warm and root growth rapid. The quicker the planting the better, and rather than wait for ideal conditions, planting may be done under surface conditions that may not be ideal without detriment to land or plants, except in the very high rainfall areas. From late October onwards, the soil will be so cold that root growth will be slow and delaying planting to obtain better conditions is wise.

The soil should be friable so that it settles back around the roots satisfactorily; hence the reason for not over-consolidating the land prior to planting. Planting should not be done under wet or frosty conditions. In the spring the strawberry starts making new root early and planting done as soon as possible, even if not under ideal conditions, is better than waiting and losing valuable weeks of root growth in the permanent quarters.

Planting by Hand

A hand trowel, dibber, spade, or in certain areas a long-handled shovel are used. The most frequent problem is to make a hole of sufficient depth to avoid doubling of the roots. This is the reason for the practice of trimming the roots of very large runners.

On soils liable to smear, a dibber or spade may produce a most unsatisfactory hole, the sides of which may be impenetrable to the roots, and trowel planting is better. But dibber planting in good fertile soil, with plenty of humus in it, is faster than trowel planting.

The runner should be planted so that the crown or rootstock is level with the soil. If planted too deeply the crown may get covered with soil, the leaves then have difficulty in emerging and the runner may die. If planted too high the roots are partially exposed and new roots emerging from the base of the crown cannot establish readily. Plants with exposed roots are more liable to herbicide damage.

The runner should be placed in the prepared hole at the right height and the soil firmed back with the implement. In loose soil the foot may be used to firm the runner, particularly if the soil is dry, but this is unwise under wet conditions. An experienced planter, if the plants are dropped ahead in position down the row, should plant about 200–300 runners per hour.

Planting by Machine

Several planting machines are available that are suitable for strawberries. The simplest types are those with two coulters which open a furrow and the operator, sitting on the machine, bends down and holds the runner in the open furrow. Two press wheels—partially weighed down by the weight of the operator—press the soil as it falls back around the plant. The depth that the runner is planted and the distance between plants depends entirely on the skill of the operator.

More expensive planting machines use snap, or rubber-coated grip, fingers, or spring loaded discs to hold runners, which are then carried down behind the coulter and released as the press wheels firm the soil around the plant. Again, depth of planting depends not only on the adjustment of the machine, but on the accurate placing of the runners within the fingers or discs of the machine.

Provided soil conditions are good, the machine accurately adjusted and the tractor makes steady progress over even ground, machine planting can be better and more uniform than hand planting. Under dry conditions, machine planting is generally better than by hand, since moist soil is folded back around the roots. Plants with very extensive root systems are difficult to plant by machine—as by hand—and the roots may get doubled. Under these conditions root trimming to the length possible to plant is necessary.

Not all machines will travel slowly enough for very close plant spacing; the speed at which the operator can work is too slow for the tractor speed. A tractor with a reduction gearbox is generally necessary. The row distances may also be limited by the spacing of the planting units—usually two—and the tractor wheels. Care should be taken not to run the tractor wheel where the next row is to be planted. However, machine planting is so much faster than by hand that it may be preferable to fit the plant rows to the tractor wheels and planter, and if necessary compensate by altering the in-row spacing between plants. Machine planting should not be attempted under wet and sticky conditions when traction is difficult, and it is under these conditions that hand planting may still be necessary.

A planting machine with two planters, supplied with properly graded plants well laid into the trays in front of the planters, should plant about 2000 plants per hour at 0.9×0.5 m spacing. Where a grower is lifting his own runners, machine planting enables him to lift during the morning and early afternoon and plant in the afternoon, thus ensuring cooler night conditions on the newly transplanted runners. This is particularly valuable during hot weather in August or September.

Cold stored runners are sometimes difficult to plant by machine because there are few or no leaves by which the plant may be gripped. But if care is taken the crown can be held and good results obtained.

Deblossoming

Deblossoming is the removal of the flower truss; it is done to prevent the plant fruiting, as is necessary in runner beds. Deblossoming should be done when the truss has lengthened, using a knife or scissors. The truss should not be pulled off or the crown may be damaged. If left until the flowers have set and the fruit is swelling, deblossoming will be less beneficial to the plant.

Cold stored runners must be deblossomed in the autumn after planting, unless they have been planted very early and have made strong growth. The fruit produced from runners that have not been deblossomed is generally worthless and will depress flower initiation and hence the crop the next summer (page 34). Late winter or spring planted, freshly dug runners will also benefit by deblossoming in the spring, if labour is available, but commercially this is seldom done; the small amount of fruit produced is generally left unpicked.

Strawing

Strawing must not be done too early as the risk of frost damage to flowers and fruit is greater once the straw is laid. The right time is just before the developing fruit begins to weigh down the truss to the ground.

The straw should be clean and free from weed seeds. Barley straw is preferred as it is more pliable than wheat straw and can be more easily tucked around the plants, under the fruit trusses. Oat straw may be a source of stem and bulb eelworm, and should not be used. About 2.5 tonnes/ha is usually required; one-year plants that are fruiting can be efficiently strawed with less. On PYO holdings plenty of straw should be used to give a good clean appearance to the customer.

Straw nowadays is baled into approximately 25 kg bales which can be carted into the field and down the rows on a pallet transporter or trailer, and can be dropped as required. One operator opens the bale and spreads the straw roughly between the rows. Other workers follow on hands and knees, and lay the straw carefully under the fruit trusses, lifting the leaves and trusses with the flat of one hand and tucking the straw in with the other. Matted rows are often not strawed and if so, only in the alleys. On very stony or coarse sandy soils it is not always necessary to straw. Fine sandy, silty loams are the most tenacious and more inclined to stick to the berries after rain. When herbicides are used and the soil surface is not disturbed, strawing may not be necessary, with saving in costs.

Some growers have made strawing machines. These consist of various devices to chop the straw and deposit, or blow it, into the alleyways. The bales are carried on a platform behind the tractor and two or three operators feed the bales into the machine. Most strawing machines will give a good layer of straw down the alleyways but not work it below the fruit trusses. A certain amount of hand labour may be used, after a machine, to push the straw under the plants. At present most strawing is still done by hand and the job is often useful to gather pickers together before picking begins. It is usual to allow 90 hours of labour/ha to lay the straw, plus labour for carting.

Control of Runners

If a matted row is being grown, the runners should be allowed to root along the rows until sufficient crowns are thus obtained. Runners can be brushed into position by workers moving quickly along the rows using short sticks; this must be done before the runners are rooted. Alternatively, light weeding attachments on tractors can be set to pull the runners into place, but this may cause soil disturbance, undesirable where the ground has been treated with herbicides. Excess runners not required in the matted row should be removed and where spaced plants are grown all runners must be removed.

This operation is difficult to mechanise completely. Rotary cultivators are frequently used in the alleys to destroy runners, but runners in the spaced plant rows need brushing out by hand before rooting, or if rooted, need chopping out with hand hoes, or cutting off with knives or hand sickles. Runners should not be pulled off as they may break the branch crowns of the parent plant.

Some chemical control of runners can be achieved with the desiccant paraquat used at the rate of 6 l/ha (4pt/acre) in sufficient volume of water. The finer the mist spray the better the killing effect, but unfortunately the greater the drift and the possible damage to the parent plants. The methods of application are detailed in the chapter on 'Weed Control with Herbicides and Cultivations'. Adjustable vertical guards to shield the row of plants from the spread of paraquat in the alleys are well worthwhile and may be made from two disc coulters which also sever the stolons and prevent translocation of the paraquat to the parent plant.

Paraquat is most effective on dull, cool days; hot sunny days reduce its efficiency. It will also effectively kill annual weeds. Runner control using paraquat works well in summer and early autumn, but rooted runners in winter and spring are more resistant and may need several applications.

There may be some translocation of paraquat from the runner to the parent plant, causing yellowing of the young leaves in the plant centre. Cambridge Favourite is particularly susceptible, but it does not appear to reduce the crop. Translocation takes place most noticeably in the spring. Direct paraquat damage will show on plants as brown dead spots on the leaves with some chlorosis. Chemical means of inhibiting the production of runners have not yet shown promise.

Cleaning after Harvesting and Defoliation

After picking some fields are often weedy, particularly if the herbicide programme has been unsuccessful and the season has been wet. It is important to clean the beds immediately after harvest since the plants rapidly make new roots then and are initiating flowers for the next crop. Delay in removal of weed competition can reduce the potential of the crowns for next year's crop.

On vigorous plants prompt removal of the foliage by mowing off all the leaves with a hay cutter bar, rotary cutter or flail harvester, or on a smaller scale using a hand rotary scythe, or shears and sickle, generally improves the number of trusses initiated by the plant and the crop the following year. However, if this mowing is delayed for some weeks after harvest, the crop will be reduced and also, if the mowing is followed by very hot dry weather, the plants may be debilitated. Weak plants should not be mown.

In general the hygienic value of rapid removal of the old leaves, particularly on vigorous plants, outweighs any disadvantage that may occasionally occur. The implement should be set at least 10 cm from the ground. This mowing will also sever some of the runners. The leaves, weeds and straw after a few days wilting may need removal, either by burning or they may be flailed and blown into a silage trailer and carted off.

Provided there is a good brisk wind going up or down, not across the rows, burning is an excellent method. If the straw has become much entangled with the runners and trodden into the soil during a wet picking season, it is difficult to burn unless it can first be scratched up by a tractor mounted cultivator, or by hand forks. If by hand, it is well worth while to toss some on the row to encourage a good fire along the crowns. Although late July and August are often wet times, there are generally periods of dry weather when burning can succeed. Burning reduces the pests and diseases from old plant debris in next year's crop, and leaves an excellent, clean surface ideal for herbicide applications. It is important to get quick burning down the rows, hence the need for wind. In hot dry summers a slow-burning fire may kill out the crowns of the plants. Weak plants and those suffering from red core or verticillium wilt must not be burnt.

Plants grown in matted rows are generally not mown or burnt. Sometimes a rotary cultivator with the blades spaced out is used to turn in the straw, weeds and runners, leaving about 0.5 m wide strips containing the original plants and the self-set runners.

Many plantations are not mown or burnt; the straw is left on the soil surface and added to next year. Mowing is particularly of value on spaced vigorous plants, or where there are a lot of weeds as the plantation is left in a more manageable condition. If a flail is used carefully to mow off the leaves the debris can be blown into a trailer—as if for silage—and removed and the straw left between the rows. Whether to mow and burn, to mow and clear, or not to mow, or to use a rotary hoe, must be decided for each plantation, each year.

After the field has been burnt over, or rotary cultivated, the soil should be clean enough for the immediate application of the herbicide simazine. If the soil is very dry, irrigation before the herbicide is applied can help in the regrowth of the plants and also to activate the herbicide. Fertilisers are best applied to older plants after they have been cleaned and, to avoid damage to the new leaves, before regrowth commences.

Ideally, thorough subsoiling should be done before the plantation is planted and it will then not be necessary during the remainder of the life of the crop but can be done again when the field is grubbed. However the passage of spray machinery, certain weather conditions, or inadequate pre-planting soil preparation may indicate that subsoiling down the strawberry rows would be beneficial. After harvest, when the soil is often dry, is a good time to subsoil a plantation. If done immediately after harvest, it helps root growth which starts to be active again at this time. Subsoiling amongst plants should not be done under wet conditions or during the winter.

Irrigation

The drought summers of 1975 and 1976 showed the importance of deep, moisture retentive soils and healthy plants with good, deep root systems for high yielding strawberries. Plantations on light sandy or gravelly soils, or those suffering from root disease of one form or another, rapidly dried out in those years, and the crop was not worth picking.

Providing the strawberry is grown on moisture retentive soils of good depth, adequate crops can usually be picked under English weather conditions in June and July without the need for irrigation. In fact on these soils irrigation may be more valuable to activate herbicides, or to aid early autumn planting, rather than cropping. Extravagant use of irrigation may actually be detrimental, encouraging growth of leaves and stolons at the expense of fruit and flowers and increasing the incidence of botrytis rot.

On the lighter soils irrigation carefully applied just before, or during the picking season will be beneficial and generally increase total yield and individual berry size. Irrigation is unlikely to be required before or during blossoming; even in dry years there are sufficient soil moisture reserves at this time.

To apply irrigation sensibly with due economy of water, and to apply it when it will do most good and is really needed by the strawberry plants, requires some knowledge of the soil moisture reserves in the rooting depth of the plantation, the rainfall, and the potential transpiration for the area. This latter can be obtained from the local ADAS office from information supplied by the Meteorological Office, but local rainfall is best measured by installing a gauge on the holding.

The reserves of moisture in the soil depend on soil type, texture and depth. In good soils strawberries will root to one metre, but in less fertile soils most of the roots will be in the top 40 cm depth. Light sandy and gravelly soils hold about 40 mm of moisture per 30 cm depth; medium loams, sandy loams, silts and clays between 40 and 60 mm per 30 cm depth of soil and peats and very fine sandy loams can hold up to and over 60 mm of soil moisture per 30 cm depth of soil. Depending on the depth of rooting some idea of the soil moisture available to the crop is thus known.

A simple irrigation balance sheet of soil moisture deficit (SMD) can be kept on a weekly basis. The balance sheet should start in April, as in most years the soil is at field capacity then, that is holding the maximum of soil moisture. As the weeks go by this soil moisture is depleted by the potential transpiration less any rainfall or irrigation applied. In some weeks there may be a loss of soil moisture, in some weeks after irrigation or rainfall there could be a gain. The running total in millimetres is a guide to the calculated amount of water that would have to be applied to restore the soil to field capacity, that is the SMD.

A rough guide to irrigation of strawberries growing on light soils would be to apply 25 mm at 25 mm SMD. If this resulted in too frequent application apply either 25 mm at 50 mm SMD or less frequently 40 mm at 70 mm SMD or above; this latter method would be satisfactory for the loamy and peaty soils. If rooting depth is good, the SMD can be allowed to increase up to 100 mm and up to 50 mm irrigation could be applied less frequently.

All these measurements are of course guidelines. The important point is to measure the amount of water applied by irrigation, either through a water flow meter, or less satisfactory, by using cans to catch the irrigation in the crop to check what has been applied. In practice irrigation has to be a compromise between what is theoretically ideal, the amount of water available and whether labour and equipment are available to apply it and the crop in a suitable condition. Irrigation should be timed carefully and preferably done before, rather than after, spraying fungicides. It is always best to irrigate immediately after, rather than before picking; ripe berries can be injured by heavy irrigation.

As the strawberry is a ground covering plant some of the self-propelled rainguns used for agricultural crops may be safely used on strawberries. These require less labour than portable overland pipes with sprinklers which are more often used. Mobile overhead spray gantrys are also successfully used to irrigate strawberries provided the field is reasonably level.

Water is now controlled by the Regional Water Authorities and a licence is required for abstraction and use. Full details of irrigation equipment, water storage and distribution and the licensing requirements are given in the HMSO publication, *Irrigation.* Advice can be obtained from the Drainage and Water Supply Officers of the Ministry at Divisional Offices.

Weed Control with Herbicides and Cultivations

Before the development of herbicides weed control operations were inseparable from those causing soil disturbance and the highly beneficial effect of keeping the crop free from weed competition may have implied that cultivations themselves were necessary operations. Numerous trials have shown that in the absence of weeds the strawberry grows and yields as well, or better, with non-cultivation as with cultivation.

Weed control in strawberries was always an expensive item and weeds often became unmanageable, especially during wet weather and during the long period from strawing until after picking. For these reasons many growers have now adopted systems of non-cultivation using herbicides to control the weeds; others use a combination of some cultivation after picking and in the autumn to turn in the plant debris and remove runners, and then use herbicides in late autumn and spring.

If the ground is very uneven after planting, it may be wise to cultivate very lightly. Soil left in large clods after planting and then treated only with herbicides is too rough for strawing down and picking, and the weed control may be unsatisfactory.

The chemicals available for weed control are of two types:

residual herbicides, applied to the soil;
contact herbicides, applied to the growing weeds.

Herbicides that have been approved under the Agricultural Chemicals Approval Scheme carry the approval mark on the label (page 84). The Booklet HSF 21, Chemical Weed Control in Strawberries should be consulted as it is revised more frequently than this publication. Use of some residual herbicides is related to soil types (Table 3).

There are three clearly defined stages for the use of herbicides in strawberry growing:

1. Perennial weed control before planting, but also in the crop.
2. Annual weed control at planting, or in young plants.
3. Annual weed control in the established crop.

The common names used for the weeds mentioned here are according to the recommended list sponsored by the Botanical Society of the British Isles, entitled 'English Names of Wild Flowers'. This book published by the Butterworth Group has been produced to overcome the ambiguity caused by referring to the same plant species by different colloquial names.

Perennial Weed Control

Frequent references have been made in this publication for the need to ensure weed free, thoroughly prepared land prior to planting, because once the crop is planted deep rooted perennial weeds cannot be effectively controlled without risk of crop damage. The weeds most likely to be present in the field will be influenced by the past cropping. Knowledge of this is useful in planning the preparation of the field prior to planting. Bindweed, and to a lesser extent thistles, are suppressed rather than killed by the dose of the hormone type herbicides used in cereals. Similarly the presence of a grass sward inhibits the aerial growth of thistle and other weeds, without killing the root system. Therefore, when strawberries are following any crops other than those in a vegetable rotation, it should not be assumed that perennial weeds are not going to be a problem, simply because they cannot be seen.

Perennial, and difficult annual, weeds likely to follow particular cropping are:

Old pasture – creeping thistle
Cereal cropping – common couch, creeping thistle, knotgrass and speedwells
Ley which has followed cereals – common couch
Grubbed top fruit in grass – docks, creeping thistle
Top fruit plantations where a herbicide strip has been maintained – bindweed
Vegetable cropping – speedwells

Whatever the preceding crop, it is best to allow at least a full cropping season in which to clean the land of these difficult weeds.

Control Measures prior to Planting in a Fallow

Traditionally a bare fallow with frequent cultivations has been the method used for weed control before planting. When properly done in a dry season, shallow rooting perennials such as common couch may be controlled. Deeper rooting perennials such as thistle and bindweed will only be suppressed, to reappear after planting. The best method is to combine a bare fallow with foliar applied translocated herbicides. A residual herbicide such as simazine should first be applied in the early spring to control annual weeds. The perennial weeds which appear must then be allowed to form a good canopy of foliage. Glyphosate is a very efficient translocated herbicide, controlling most perennial weeds, as well as all annual weeds present at the time of spraying. There is no need to plough after this herbicide. Alternatively aminotriazole or dalapon will give a good control of common couch, whilst 2,4-D will control bindweed, thistle and docks, though asulam is a more specific dock herbicide. Three to four weeks after these herbicides it is necessary to follow with deep ploughing. Strawberries must not be planted for two months after aminotriazole or four months after dalapon applications.

Control of Perennial Weeds in the Newly Planted Crop.

However efficiently the pre-planting treatments have been carried out, perennial weed seeds can still germinate after planting. It is essential to control these weeds in the first year to prevent a serious infestation developing. The herbicides suitable for use immediately after planting strawberries (Table 4) can also control many germinating perennial weeds. Once past the seedling stage spot application with a suitable herbicide, or hand pulling and hoeing are the only remedies.

Control of Perennial Weeds in the Established Plantation

Propyzamide may be used between October and December to control established common couch. It will also give complete control of the germinating stage, but only partial control of the seedling and established stages, of creeping buttercup, docks and sheep's sorrel. With this herbicide there are restrictions as to dose and soil types and use on matted rows. Terbacil can severely damage strawberry plants but may be carefully used as a spot treatment for the control of couch grass. Some control of docks, bindweed, thistles and perennial nettles may also be achieved.

2,4-D amine may be applied immediately after fruiting to control thistles, bindweed, nettles, docks and dandelions, though some crop damage may occur. It is important, when using 2,4-D amine, to spray before any mowing since the old leaves protect the crowns and prevent the herbicide affecting flower formation. Defoliation may be done one week after spraying. Upright free standing weeds such as thistles, docks and nettles can be treated with glyphosate* applied through a herbicide glove when the weeds are in full growth. This is much safer than terbacil or 2,4-D amine applied through a sprayer but requires much more labour. Prostrate weeds such as bindweed, clover and common couch can safely be controlled by this method.

Table 2

Herbicides for perennial weed control in strawberries

Timing	Weed	Chemical
Before planting	couch and broad-leaved weeds	glyphosate, aminotriazole
	docks only	glyphosate, asulam
	broad-leaved weeds only	glyphosate, 2,4-D
	common couch only	glyphosate, dalapon
After planting strawberries (overall or spot application)	broad-leaved weeds	2,4-D
	couch and dock seedlings	propyzamideϕ
	clover	ethofumesateτ

ϕ only on plants established for at least one year. Not on matted rows

τ only on plants of cv. Cambridge Favourite that have been established for at least one season

Timing	Weed	Chemical
After planting strawberries (spot application only)	free standing weeds	glyphosate*
	common couch only	terbacil

*via Croptex Herbicide glove only

*This recommendation has now been withdrawn by the manufacturers of glyphosate

Table 3

ADAS classification of soil textures for use with soil acting herbicides

Textural class	Symbol	Textural group
Coarse sand	CS	
Sand	S	
Fine sands	FS	Sands
Very fine sand	VFS	
Loamy coarse sand	LCS	
Loamy sand	LS	
Loamy fine sand	LFS	Very light soils
Coarse sandy loam	CSL	
Loamy very fine sand	LVFS	
Sandy loam	SL	Light soils
Fine sandy loam	FSL	
Very fine sandy loam	VFSL	
Silty loam	Zyl	
Loam	L	Medium soils
Sandy clay loam	SCL	
Clay loam	CL	
Silt loam	ZL	Heavy soils
Silty clay loam	ZyCL	
Sandy clay	SC	
Clay	C	Very heavy soils
Silty clay	ZyC	

Notes

1. Adsorptive capacity for herbicides increases with soil organic matter content.
2. Organic matter content tends to increase with clay content, 'heavy' soils generally contain 2.8-4.0 per cent organic matter compared with 1.0-1.5 per cent on 'very light' soils.
3. Within each textural group the organic matter content will be lowest under continuous arable cropping, particularly in the low rainfall areas of southern England and highest in ley-arable systems, in the high rainfall areas of the north and west. Thus optimum doses for a textural group may be slightly higher in the north and west than in the south and east of the country.
4. The prefix 'organic' is applied to the above mineral texture classes if organic matter levels are relatively high and greater than ten per cent. The 'sands' and 'very light' soils, however, appear 'organic' when they only contain five to six per cent organic matter.
5. The prefix 'peaty' is applied when organic matter levels are between 20 and 35 per cent.
6. 'Peat' soils are those containing more than 35 per cent of organic matter.
7. The risk of herbicide leaching on very stony or gravelly soils is greater and textural grouping for such soils should be modified accordingly.
8. Highly calcareous soils appear very 'silty' in nature and may well behave more like 'light' rather than 'medium' soils towards soil acting herbicides.

Annual Weed Control at Planting or in Young Plantations (Table 4)

Residual Herbicides before Planting

Under dry soil conditions and where irrigation is not available and particularly if knotgrass, annual grasses and fat-hen are likely to be troublesome weeds, the use of trifluralin before planting by incorporation into the soil should be considered. It is also safe before planting cold stored runners in mid-summer, which are always irrigated (page 34).

Residual Herbicides after Planting

Several herbicides may be used, the choice depending on time of planting, soil type, expected weeds and whether or not irrigation is available. Of the herbicides available lenacil is the most widely used (Table 4), although the tank mix of propachlor plus chlorthal-dimethyl is probably safer on Very Light Sands. Table 3 describes the ADAS classification of soil types.

Three useful herbicides that can be safely used on plants established for one year should not normally be used on newly planted runners particularly on Light soils because they cause damage or death; these herbicides are simazine, trietazine plus simazine and propyzamide. But coating the roots with activated charcoal, either as a powder or a slurry before planting protects the plants and allows the post-planting use of simazine or trietazine plus simazine at the grower's risk.

Table 4

Some herbicides for use at planting for annual weed control

Name	Limitations to use: Timing	Limitations to use: Soils
Pre-planting residual (incorporated)		
trifluralin	0–14 days before planting	not on Coarse Sands or Sandy soils. Not on soils containing more than 10% organic matter
chlorthal-dimethyl	no limitations	dose varies with soil type. Do not use on organic soils
propachlor	no limitations	do not use on organic soils
chlorthal-dimethyl/solidus propachlor	no limitations	dose varies with soil type. Do not use on organic soils
Pre-planting residual (surface applied)		
chlorpropham/ fenuron	10–14 days before planting	formulations and recommendations vary. See manufacturers literature for correct choice according to soil type.

Post-planting residuals

chlorthal-dimethyl	no limitations	dose varies with soil type. Do not use on organic soils
propachlor	no limitations	do not use on organic soils.
chlorthal-dimethyl/ propachlor	no limitations	dose varies with soil type. Do not use on organic soils
lenacil	no limitations	Very Light and Light Soils Silts, Warps and Medium Loams. Note, do not use on Coarse, Very Fine, Sandy or Gravelly soils or soils containing more than 5 per cent organic matter
2,4-DES	no limitations	no limitations
chloroxuron	14 days after planting	Light soils, lower dose Heavy soils, higher dose
2,4-DES/ simazine	14–21 days after planting	no limitations
diphenamid	26 weeks after planting (not before new foliage has appeared)	Light soils, lower dose Heavy soils, high dose
simazine	in November–December following early autumn planting provided 3 new leaves have formed	no limitations

Contact Herbicides before Planting

Paraquat is non-selective and will destroy all green matter in the presence of light but is inactivated on contact with the soil. It will kill annual weed seedlings, though knotgrass and cleavers tend to be resistant: these may however be killed at the two to three leaf stage by increasing the quantity of wetter in the spray. Paraquat may be used to control annual weeds which have germinated between the final preparation of the ground and planting, and so provide a weed-free soil for the application of post-planting residual herbicides. This is a particularly valuable technique when planting in August and September. Paraquat may also be used post-planting as a directed application between the rows, when low pressure sprays or non-drift type nozzles must be used to prevent the chemical falling on the crop.

Contact Herbicides in the Crop

Phenmedipham will control common chickweed, groundsel and speedwells at the seedling stage, but annual meadow-grass is resistant. Phenmedipham may cause a check to growth and some yellowing of leaves but this should not be serious and plants soon recover. Barban may be added as a tank mix for improved control of black bindweed, fat-hen, knotgrass and redshank seedlings and in addition the control of wild-oats. Whether used alone, or in a mixture with barban, the dose of phenmedipham should be reduced when temperatures at spraying are 21°C or higher, and spraying delayed until after 1700 hours.

Annual Weed Control in Established Plants (Table 5)

No herbicide, residual or contact, should be used in cropping beds between flowering and harvesting.

Residual Herbicides

Lenacil can be applied to strawberries at any time provided that no further application of any residual herbicides is made within three months. The same limitations on soil type apply as for the use immediately after planting. Because of its ineffectiveness on dry soils it is mainly applied in February, when the soil is moist. Following the early February application, a second application at three quarters dose may be made just before flowering in late April to early May, to extend the residual life through the cropping period. Although the time interval is slightly less than three months, commercial experience shows it to be safe, and particularly worthwhile in the first year.

Simazine should only be applied when the strawberry crop has been established for six to nine months, according to planting date, and only during the period July to December. The recommended dose is often split, half being applied in July/August after harvest and half in November/December. Simazine is relatively long lasting, and from winter application should give effective control until harvesting, but it does not control knotgrass or speedwells.

Chloroxuron may be used at any time after picking until (a) six weeks before strawing or (b) not later than mid-April or (c) the commencement of flowering, whichever occurs earliest. As with the post-planting treatment the dose varies according to soil type.

Propyzamide should not be applied to strawberries established for less than twelve months. Its use is limited to one application within nine months and between October and December. It can only be used on Loams and heavier soils containing less than ten per cent organic matter. At the recommended lower dose established annual grasses, volunteer cereals and common chickweed will be controlled, together with seedlings up to the two-leaf stage of fat-hen, small nettle, black bindweed, black nightshade, knotgrass, redshank and speedwells. To control established plants of these weeds, together with seedling cleavers, common fumitory and shepherd's-purse, the dose must be increased, but this higher dose is not recommended for use on matted row strawberry crops. Despite these limitations, its ability to control established plants of annual grasses and common chickweed, which are resistant to lenacil and simazine, makes propyzamide a very useful herbicide. It is rapidly broken down in warm weather, and effectiveness may consequently be severely limited.

Trietazine/simazine may be applied to crops planted for at least one year, the dose varying with soil type. The mixture is cheaper than lenacil, yet offers control of speedwell and field pansy, whilst it has the advantage over simazine of controlling knotgrass. It may be applied in the autumn or spring.

Chlorthal-dimethyl or propachlor separately or as a tank mix may be applied at any time other than during the flowering or fruiting periods. Chlorthal-dimethyl is especially valuable for the control of speedwells, either pre- or post-emergence, whilst propachlor offers good pre-emergence control of cleavers.

Table 5

Herbicides for use in the established crop for annual weed control

Name	Limitations in use Timing	 Soils
Residuals		
lenacil	i. at any time outside the flowering to harvest period. ii. no more than 2 applications, per year. iii. at least 3 months should elapse before another application of any residual herbicide is made.	as for use at planting
simazine	i. July–December ii. either as a single dose or as a split dose, part in July–Sept; and part in Nov.–Dec. iii. with no more than half the total dose going on in Nov.–Dec.	i. not on organic soils ii. use with caution on Very Light or Gravelly soils. iii. on other soils use lower dose and high dose on Medium/Heavy soils.
propyzamide	i. only to plants established one year. ii. apply between Oct.—Dec. iii. only 1 application in any 9 month period. iv. Simazine must not be used within 6 months, before or after the use of propyzamide	i. not on soils of more than 10% organic matter. ii. not on Sands or Very Light soils
trietazine/ simazine	apply to plants when established for one year.	applicable to all mineral soils, dose varies with soil type
chlorpropham/ fenuron	mid-Sept.–mid-Dec.	as for pre-planting use

chloroxuron	i. spring use: apply (a) 6–8 weeks before strawing, (b) not later than mid-April ii. autumn use: apply immediately after post-harvest clean up iii. Cloche production: (a) 2 weeks before covering (b) rain or irrigation desirable between application and covering iv. runner beds: (a) just before onset of running (b) rain or irrigation should follow application	lower dose on light soils; higher on heavy soils.
2,4-DES/ simazine	i. apply after harvest ii. October–January iii. not more than 3 applications per year	no limitations to soil type
2,4-DES	i. apply after harvest ii. October–January	soil must be moist no limitations to soil type
chlorthal-dimethyl	no limitation outside the flowering-fruiting period	dose varies with soil type. Do not use on organic soil
propachlor	no limitation outside the flowering-fruiting period	do not use on organic soil
chlorthal-dimethyl/ propachlor	no limitation outside the flowering-fruiting period	dose varies with soil type. Do not use on organic soil
diphenamid	apply i. when dormant or semi-dormant ii. after harvest	Light soils, low dose Heavy soils, high dose
Post-emergence contact		
phenmedipham	at any stage except from beginning of flowering to picking	no limitations
phenmedipham/ barban	at any stage except from beginning of flowering to picking	no limitations

Chlorpropham plus fenuron mixtures may be used at any time between mid-September and mid-December, the choice of product being dependent on soil type. Diphenamid at the dose given for the post-planting treatment, may also be applied to the established crop either immediately after harvest or in the dormant period.

Table 6

Annual Weed Susceptibility to some herbicides

S=susceptible MS=moderately susceptible R=resistant MR=moderately resistant

Chemicals and timing	Residual established crop														Foliage Acting	
		pre-planting			post-planting											
Weeds	chlorpropham+fenuron	trifluralin	chlorthal-dimethyl	propachlor	propachlor+chlorthal-dimethyl	2,4-DES	lenacil	chloroxuron	2,4-DES+simazine	2,4-DES+chlorpropham	diphenamid	simazine	trietazine+simazine	propyzamide	phenmedipham	phenmedipham+barban
Annual meadow-grass	S	S	S	S	S	R	S	MS	S	S	S	S	MS	S*	R	R
Black-bindweed	S	S	–	MR	MS	MS	S	MR	MS	S	–	MS	MS	S‡	S	S
Black nightshade	MS	R	S	MS	S	MS	R	S	S	MS	R	S	MS	S‡	MR	MR
Charlock	MS	R	–	R	R	S	S	S	S	S	–	S	S	R	MR	S
Cleavers	MS	R	–	S	S	R	R	R	R	R	–	R	R	S	MR	MR
Common chickweed	S	S	S	MS	S	S	S	S	S	S	S	S	S	S*	S	S
Common fumitory	MS	S	–	R	R	MS	S	MR	S	S	–	S	MS	MS	S	S
Common hemp-nettle	MS	S	–	S	S	R	MR	S	S	MS	–	S	MS	R	S	S
Corn spurrey	S	MS	–	–	–	MR	S	S	MS	S	–	MS	S	–	S	S
Dead-nettle	R	MS	–	S	S	R	MS	MS	S	R	–	S	S	R	S	S
Fat-hen	MS	S	S	MS	S	S	S	S	S	S	MS	S	S	S‡	S	S
Field pansy	R	S	S	–	S	MR	R	–	MS	MR	–	MS	MS	R	S	S
Groundsel	MS	R	–	S	S	MR	MS	S	S	MR	S	S	S	R	S	S**
Knotgrass	S	S	S	MR	S	MR	S	MS	MR	S	MS	MR	S	S‡	MS	MR
Mayweeds	S	R	–	S	S	MR	MS	MS	S	MR	–	S	S	R	MR	S**
Redshank	S	S	–	MR	MS	MS	S	MS	MS	S	MS	MS	S	S‡	S	S
Scarlet pimpernel	MS	S	–	S	S	MS	S	S	S	MS	–	S	S	R	S	S
Shepherd's-purse	S	R	–	S	S	S	S	S	S	S	S	R	S	S	S	S
Small nettle	S	S	S	S	S	S	MS	MS	S	S	S	S	S	S‡	S	S
Smooth sowthistle	R	MS	–	MS	MS	S	S	MS	S	S	–	S	S	R	R	R
Speedwells	MR	S	S	S	S	MS	MR	MS	MS	MS	–	MR	S	S‡	S	S
Volunteer cereals	S†	S	–	–	R	R	S	MR	S	S	S	S	–	S*	R	R
Wild-oat	S†	MS†	–	–	R	R	R	R	S	S	S	S	R	S*	R	S

†only those germinating near the surface (in the herbicide layer) will be controlled

‡also controlled post-emergence to the seedling stage

*also controlled post-emergence to the established plant stage

**at one true leaf stage

Since printing 2.4-DES/chlorpropham has been withdrawn from the market

2,4-DES and 2,4-DES plus simazine can be applied after harvest, and between the period October to January, but where the simazine mixture is used, no more than three applications of the recommended dose may be made in one year. The addition of chlorpropham to 2,4-DES limits its use to the period October to December only.

Contact Herbicides

Paraquat in addition to controlling emerged annual weeds at planting, may be used in a similar manner at any time in the life of the crop. Phenmedipham as used at planting is also applicable to the established crop. The chemical is most active under warm conditions and this may influence its effectiveness when used in early spring to control germinated weeds resistant to simazine and lenacil. Satisfactory control may not be achieved under cold conditions. A phenmedipham/barban mixture can also be used.

Runner Control

The contact herbicide paraquat can be used for runner control. Two applications are generally necessary, one in August and the other in October/November. There is however, a risk of damage to the parent plant from the October/November application if the August application is omitted. This is due to translocation of the chemical from the full stand of runners. This risk can be reduced by separating the runners from the parent plant using a slicing disc down the sides of the rows.

Application of Herbicides

Having decided on suitable herbicide treatments, it is equally important to ensure even and accurate application. The volume of water applied is not critical, provided it is sufficient to give an even ground cover. Residual herbicides are normally applied in 500–1000 litres/ha (50–100 gal/acre), the higher rates being preferable for small-scale operations, while lower rates may be satisfactory with larger, more accurate sprayers. Manufacturers' instructions should be followed when mixing, and thereafter continuous agitation is essential to avoid herbicides settling out in the spray tank.

Most residual herbicides are used as overall treatments, and it is important to avoid overlapping of sprayed areas particularly on the plant rows which would result in overdosing and crop damage.

Spraying Equipment for Applying Herbicides

The type of equipment used depends primarily upon the scale of the operation, but in each case the height of the nozzles, constant spray pressure and uniform speed of travel are important for accurate applications.

Tractor-mounted herbicide sprayers are operated from the power take-off, with the tank and booms mounted on the three-point linkage of the hydraulic system. Normally these apply about 250 litres/ha (25 gal/acre) but for efficient application it is better to use 500–1000 litres/ha (50–100 gal/acre). Either ceramic fan nozzles or disc nozzles are suitable, but their pattern of distribution differs. Fan nozzles give a more even spread, but more drift in windy weather than

Strawberry plants, variety Cambridge Favourite showing production of stolons and runners

Strawberry runners as removed from parent plant showing range of size. The very small runners are only suitable for very early planting with irrigation

PLATE I

Timing of sprays for *Botrytis* control on Cambridge Favourite
a. white bud, b. first flower, c. full flower, d. petal fall.

Photos: Long Ashton Research Station.
Anther failure in Redgauntlet

a. Most anthers healthy, yellow and some shedding pollen.

b. Small unhealthy anthers, dull-brown and without pollen.

A ruffler chain suspended between the two spray arms with nozzles aids spray penetration of dense plants.

PLATE III

Photo: A G D Heath ARPS
All dessert fruit must be carefully picked by the stalk.

Straddle mechanical harvester at work in strawberry plantation.

Photo: A G D Heath ARPS

Good sample of plugged Cambridge Favourite in processor's tray.

Photo: Miss H M Hughes

Patch of couch grass and other perennial weeds regrowing in spite of herbicide treatments that have damaged the strawberry plants.

PLATE V

Early strawberries grown on ridges, protected by low continuous tunnels.
Tunnels opened to aid pollination.

Photo: A G D Heath ARPS

Film plastic tunnels are used to produce early fruit. If Redgauntlet is grown, the plants produce secondary trusses to give late summer and autumn ripening.

Arabis mosaic virus in Cambridge Favourite.
Note flecking and mottling of some leaves.

Photo: A G D Heath ARPS

If protectant sprays are not used *Botrytis* fruit rots usually develop.

Red core symptoms seen when infected roots are cut lengthways.

Red spider mite infestation on Cambridge Favourite.
Note speckling, bronzing and distortion of leaves.

Stem nematode infestation on Cambridge Favourite plant showing typical leaf crinkling.

PLATE VIII

disc nozzles. The correct boom height is less critical for fan nozzles than with discs. The latter need checking for wear, while ceramic fan jets become inaccurate when chipped. Control droplet application (CDA) and ultra low volume (ULV) are generally not used for herbicide applications in strawberries. For further guidance consult MAFF Leaflets (Appendix I).

For small areas, or spot treatments, use knapsack sprayers fitted with a boom and carried on the operator's back, or on small wheels. Small machines with horizontal booms and a motorised pump which are hand pushed, or spraying attachments fitted on motorised cultivators are also available.

Special equipment is needed for some contact herbicides such as paraquat to give a directed application and avoid damage from spray drift on to the strawberry plants, unless these are being applied before planting. These contact herbicides can be applied through dribble bars, flood jets or guarded fan or disc nozzles. Contact herbicides that do not damage the crop can be applied overall with the usual sprayer boom.

Dribble bars use the flow of liquid from the tank due to gravity and operate on low pressure. Dribble bars to cover several alleyways can be tractor mounted. The dribble technique has the disadvantages that a higher volume of liquid is used, and coverage of small seedlings and runners may be incomplete, with disappointing results. Sprays are more efficient, but if used amongst the crop must be at a low pressure, using either a flood jet which produces little drift, or a well-guarded spray from a fan or disc jet. For overall treatment of land prior to planting where immediate drift is not important, a fine spray will give the best kill of weeds. Treatment with these contact herbicides must be carried out in calm weather, preferably in late afternoon or evening when several hours without rain can be expected.

Calibration of Sprayers

Tractor-mounted sprayers are provided with a chart giving rates of application at various speeds and pressures, using different sets of nozzles. A low reading speedometer and rev. counter on the tractor are essential. If information on a sprayer is not available, or thought to be unreliable, a test must be carried out to ensure accurate application.

While standing on level ground, completely fill the sprayer tank with water. Make a trial spraying of a known area while running at a suitable speed and pressure. Return to level ground and top up the tank using an accurate measure. The amount of water needed to top up to the original level can be used to calculate the rate/area. It is wise to check this figure regularly in practice.

Spraying Do's and Dont's

Do not spray if soil or wind conditions are unsuitable.
While spraying check all nozzles are delivering.
Check output of the first tankful against a known area. Stop if application rate is not correct.
Keep herbicide mixtures well agitated.
Avoid double spraying on headlands, or on plant rows.
Mark last row sprayed when leaving site for refill.
Avoid drip from nozzles on to plants; do repairs on headlands.
Empty surplus material on waste ground and wash out sprayer thoroughly.

Cultivations

Provided the soil has been adequately prepared before planting, only annual weeds will need control thereafter and light surface cultivations with A or L shaped hoes suffice. Deeper working of the soil with straight tine hoes will be necessary, if the plants are to be ridged in the autumn. This method is still adopted by some growers but has little to commend it as the ridging damages many of the surface roots. If the practice is to aid surface drainage during the winter, it is better to plant on ridges and to follow a system of non-cultivation using herbicides and leaving the ridges intact.

Hand hoeing may be necessary for control of herbicide resistant weeds. Short-handled draw hoes with sharp blades are used and workers should be trained to scrape the annual weeds off the herbicide treated ground with the minimum of soil disturbance. This will prevent the weeds seeding. Perennial weeds are best not hoed or chopped off, but dealt with by spot treatment with the appropriate herbicide at the correct growth stage. A small amount of hand labour in the first year will eliminate weeds before they have time to become established or spread widely to become a nuisance in subsequent years.

A worker will take about 120 hours to hoe the rows of a hectare of strawberries but this time will vary according to the amount of weeds and runners present. Tractor cultivations of the alleys will take about 15 hours per hectare.

The tractor mounted rotary cultivator is often used to turn in runners, weeds and straw after picking. The blades are spaced out, by removing those which run over the plant row, and it is usual to cultivate two alleys at the one time. Although the rotary cultivator will penetrate 15 to 20 cm in depth, it is best to work to the minimum depth sufficient to incorporate the material into the soil, and to increase the speed of the cultivator while travelling slowly, to ensure thorough chopping of the debris and minimum damage to the strawberry roots. Although the rotary cultivator leaves a good looking tilth a hard pan can be formed below this. Soil that has been rotary cultivated is very inclined to erode from slopes under heavy rainfall. Weeds tend to grow abundantly after rotary cultivation if not controlled by herbicides.

Subsoiling is best done before planting. There may be occasions when subsoiling between each row is done to remedy bad soil conditions.

Early Strawberries

Early strawberries can be produced by protecting plants with continuous glass cloches, or with coverings of polyethylene (polythene). Glass cloches consist of sheets of glass held in place by wires to form a simple tent, or a flat or sloping topped barn cloche over the plants. The cloches are placed end to end to cover the plant row, and the cloche at each end is generally closed with another piece of glass. Polythene film is used to form a continuous tunnel along a row of plants; the polythene is supported on wire hoops just clear of the plants and is tied down to the hoops and to a peg at each end of the row. Slightly earlier fruits are produced in film plastic houses.

A very small quantity of early strawberries is produced by specialist growers from cold Dutch light glasshouses, and also from Dutch lights on the flat. It is doubtful whether the productions during March and April of extra early strawberries from heated glasshouses is economic. The market is very limited and prices have to be high to cover the expenses of production.

The market for early strawberries, ripe during May and early June, before fruit is available from the outdoor fields, is buoyant and there is a good demand for well presented fruit at this time. In general, competitive imported fruit does not look so well as the home produce. But costs of production are high and early strawberries involve much hand labour; this may account for the popularity of protected strawberries on suitable smallholdings where the occupier can do most of the work himself, employing extra labour only for picking. Under these conditions the grower's own labour spent on detailed jobs to enhance earliness of crop should be repaid.

The area of strawberries protected with glass cloches is very small and is decreasing. The main areas are in Hampshire and the Isle of Wight, West Sussex, Kent, Somerset, Devon, Cornwall and East Sussex.

The use of polythene tunnels (Plate VI) is mainly in Hampshire, Sussex, Surrey, Kent and Somerset. This method of protection is far less costly than glass cloches but the crop is not so early. It is generally considered that glass cloches and film plastic houses will advance the crop by about four weeks, compared with field plantations of similar varieties on similar sites, and low polythene tunnels by about a fortnight. Fruit ripened under tunnels is generally cleared before main crop strawberries in the same area are ready to pick.

The prices for early strawberries on the main wholesale markets fall rapidly as supplies increase. Even so, in less favoured areas where a local outlet is assured, it may be profitable to protect strawberries and gain a few days advantage over other local suppliers. It may also be profitable to protect a few rows of strawberries on PYO holdings to help start the season.

Choice of Site

Although a light, well-drained soil that is readily warmed by the sun, is a valuable asset, climate is the overriding consideration affecting earliness. Radiation is mainly dependent on latitude, hence the advantage of the south and south-west over the midlands and north. The presence of the warm seas on the west of England and Wales also confers an advantage over areas in similar latitudes in the east of the country. The value of proximity, but not exposure to the sea, is very marked not only in direct effects on air temperatures but probably also from increased light.

Records show that the traditional early areas have higher maximum air temperatures than elsewhere. If it is assumed that temperatures above about 6°C are needed for growth, then conditions are much more favourable from January until April in the traditional early areas than in the rest of the country. Within the main overriding meteorological differences, aspect can be quite important. Calculations show that ground sloping steeply to the south can receive twice as much radiation during the winter months as level ground.

Shelter

Shelter is an important factor but must not cast shadows on the plants, particularly during the winter months when the sun is lowest. Wind is a big disadvantage to early production, since wind movements through cloches and tunnels will remove the air warmed by the radiation from the sun and trapped under the protection. The best increases of air and soil temperatures are obtained on still sunny days.

Because only small areas of plants are protected and because wind cannot only reduce temperatures but cause damage to polythene or glass, it may be worthwhile utilising artificial shelter screens to split up more exposed fields. Fences of 2 cm laths spaced 2 cm apart are ideal windbreaks. Cheaper, but less efficient, screens can be made of doubled wire netting of 1 cm mesh, or of nylon mesh netting. All these materials need well braced, strong supporting uprights, and are usually about 1 to 2 m high. Shelter screens should be permeable so that they filter and reduce but do not try to stop the wind. Solid screens, such as walls, thick woods or buildings, may create local wind turbulence and cause damage. (See HG21 Windbreaks)

No detailed experiments have been done on the best ways to site rows for earliness. On level ground the rows are usually run north to south, but on sloping or difficultly shaped fields the lines of cloches or tunnels should be positioned for ease of management.

Soil

Light soils warm more quickly than heavier soils, and well drained soils better than badly, or slow draining soils. The grower has little control over the soil once a site has been chosen, but it is probably more important to opt for good overall shelter and aspect, rather than choose a more exposed site with lighter soil. Assuming no major drainage problem exists, heavy soils can be slightly improved over the years by the incorporation of bulky organic manures, coarse graded peat, or by a system of leys. A system of growing plants on ridges may help drainage and, by increasing the soil surface exposed to the sun, increase soil temperatures.

Frost Damage

Protected strawberries flower some five to seven weeks ahead of the unprotected plants. It is unusual for early strawberries to suffer from spring frosts because they are grown in areas where there are major climatic advantages and the whole area, including the soil, is warm. However, in less favourable areas, if protected strawberries are well advanced by much sunny weather, radiation frosts could be serious, especially if they follow after a period of dull weather. Glass and polythene give protection only by retaining the heat trapped during the day and, under severe night frost conditions, polythene in particular will give little or no protection. Flowers touching the glass or polythene are most likely to be damaged.

Low Continuous Tunnels and Cloches

Planting and Selection of Runners

The planting distances will depend mainly on planting time if one-year-old plants are to be covered and to a much lesser extent on the soil and variety. Generally one-year-old plants are used; two-year-old plants are rather large and may become leafy and etiolated under cloches. Tunnels can protect two-year-old plants if these are not too vigorous. In areas where it is difficult to establish a strong one-year-old plant, a good compromise is to plant in late April, and allow some runners to root along the row, which is then covered the following spring when the plants are the equivalent of small two-year-old ones.

To obtain well grown one-year-old plants the grower may:

1. purchase freshly-dug runners in August or early September
2. raise his own plants for transplanting in August to early September.
3. use cold stored runners planted in July.

It is necessary to grow the right size of plant to fit the cloches or tunnels available and to fully utilise the ground protected. In the traditional early areas, a freshly dug runner planted the last week of August, or during the first week or so in September in the south-west, will make good root and crown growth in the autumn and give a good crop the following spring. The runners should be planted about 20–30 cm apart in single rows 0.8–1 m apart. Planted at this time, provided irrigation is available so that there is no check to growth, the size of runner has been shown to be of no importance, the planting date solely affecting total crop.

If planted after early September the growth that the transplanted runners make before the onset of winter is much reduced and runner size is more important, better growth being made by the larger runners. If planting is delayed until late September, the in-row planting distance can be reduced to 15 cm or a double staggered row used within the cloche or tunnel.

Experience will soon show the best combination of in-row distance and row distance for the protection available. Many growers plant a part of their area in late August or early September, when runners are rather expensive, and plant the remainder of the area during September when runners are cheaper and more readily available, closing up the plant spacing in the row. The earliest fruits to ripen are obtained from the smaller, late planted runners, but the best total yield from the earlier planted areas.

Row distances depend on the soil and method of covering. On sites that have been cropped with strawberries for many years and where red core disease is troublesome and vigorous growth not expected, rows are 0.8 to 0.9 m apart for low continuous tunnels and slightly wider than this if cloches are used. Sometimes a double row of cloches is used adjoining each other with a wider path to allow picking. If tunnels are used to cover two-year-old plants the rows should be at least 1 m apart. Some growers plant two single rows 30 cm apart covered by a single wider tunnel, with a path of about 1 m. As with the field strawberries, it is unwise to have very long rows without cross paths to carry out the fruit. Rows should not be longer than about 60 m.

If freshly dug runners cannot be obtained, cold stored runners have been shown to be a very suitable alternative (page 34). Where sufficient land is available, thus making earlier planting no disadvantage, cold stored runners are not only cheaper than freshly dug runners, but also can be reliably booked for delivery from store irrespective of the current season's growth of runners, or conditions for lifting. In the very warm south-west, cold stored runners of most commercial varieties can safely be planted during early August; elsewhere they must be planted before the end of July, or even much earlier in less favoured districts. If cold stored runners have made very dense and leafy growth during a mild winter, some defoliation at the time of covering in the spring is beneficial.

Use of Glass Cloches

Glass can be used in the form of tent cloches, two sheets of glass held together with a wire handle and clip under tension. Tent cloches can also be home-made, using a block of wood with a groove to take the top edges of glass at the appropriate angle. The glass is held to the wood block by a piece of wire bent round the lower edges of each sheet and fastened to a screw eye on the underside of the block. These home-made cloches are not so rigid as the purchased tension spring type. The barn type cloche is the most popular type. These cloches have four sheets of glass held together by wires under tension, and the wires permit of the removal of one of the top panes of glass to allow access to the plants. Other cloches, some with flat tops, of varying sizes and widths and with different methods of retaining the glass are available.

Cloches should go on level, clod-free soil, an important point to remember when cultivating the surface after planting and prior to herbicide applications. It is unwise to cultivate between the cloches; flying stones can cause damage. Cloches will withstand frost. The use of herbicides is advisable since then the cloches do not need to be removed for hoeing. The cloches are generally put over the plants during February. In areas where heavy snowfall or icing may occur, it is wise to wait until after the normal time for expected snowfall to avoid breakages.

In early areas, cloching during January or February or March has little effect on the time of flowering and ripening, but covering after the end of March will result in later ripening.

Over-winter covering is not normally beneficial. However, in very wet years, on soil where drainage is poor and red core disease a problem, early covering may help to shed the water into the alleys away from the plants.

It is important to place the barn type cloche with all the ventilating lights on the same side so that they can be readily opened. The ends of the cloches may be blocked with glass, wood or galvanised sheet steel cut to fit the shape of cloche and held in place with a peg and tied back to the handle of the end cloche. Not all growers trouble to seal the ends of cloche rows.

As the plants grow during April and May it will be necessary to space out the cloches and remove the ends to avoid excessive heat and etiolation and too dense growth of the plants. During blossoming an even day temperature of about 15 to 21°C is necessary with plenty of wind movement to help pollinate the flowers and care taken to avoid hot, drying conditions. The top ventilating pane of glass can be opened on to the first or second wire notch depending on weather conditions. Rows of tent cloches can be readily ventilated by removing a cloche and stacking it on its neighbour. The number to remove will depend on the growth of the plants and the weather.

For the aphicide spray in March and for fungicide sprays during blossoming, it will be necessary to move all the cloches off the plants, to the alley, or to move two cloches at the end, then spray the plants and shift the next two cloches down and so reach all the plants.

When picking commences, many growers remove the cloches and stack them conveniently in the alleys, or they may move them over an adjacent row of plants previously uncovered to get a little benefit. But in areas where birds are troublesome, or if the weather is cool so that protection from the glass is still required, the cloches are left in place and the fruit picked either by moving each cloche down the row, or by removing the top pane. Wire netting can be used over the ends of cloches to keep out birds.

Glass cloches are heavy to handle and many growers stack the cloches after use between the cropping plants on polythene. This strip method of cropping creates cultivation problems as the cloches get in the way. Trolleys, or platforms on tractor-mounted rear loaders, can also be adapted to carry cloches away from the cropping area. It is normal to find that at least ten per cent of the glass needs replacing each year. Cloches become dirty and will need occasional cleaning with warm water and detergent.

Use of Low Continuous Polythene Tunnels

This method of protection consists of continuous tunnels made of polythene (polyethylene) film 38 μm thickness suspended on wire hoops and kept down by tying to pegs at each end and to the hoops. The polythene shields the plants from the wind. It is extremely transparent to solar radiation so that all the heat from the sun penetrates the polythene and warms the plants and ground. Unfortunately, polythene also permits the loss of this heat from the soil and plants during the night. Glass, which does not allow so much back radiation through the cloche, is therefore much warmer at night. Thus during cold nights, following sunny still days, temperatures under glass cloches will be higher than those under polythene tunnels. But during sunny days, polythene tunnels, if kept closed, will often be much warmer than glass cloches where the spaces between the cloches permit the escape of warm air; sometimes the tunnels become much too hot.

Because of this difference between the properties of glass and polythene, plants growing under cloches are less liable to frost damage than plants under polythene, and this is a point that should be remembered if polythene is used to forward plants in areas where late spring frosts often occur. Under certain conditions moisture forms over the polythene film and this is invaluable as it causes the polythene film to behave more like glass and to retain more warmth at night.

SIZE OF TUNNELS

The size of wire hoop and width of polythene film to be used will depend on the size of plant expected. Tunnels can also be used to protect autumn cropping varieties and over single, or matted, rows of two-year-old plants.

It is important to assess the expected plant size as accurately as possible since the tunnel should be made to just enclose the plant. If the tunnel is too big there will be unnecessary expense and extra susceptibility to wind damage. But if the tunnels are too small the leaves, flowers and fruit will press against the polythene and will get damaged.

For runners planted at the recommended time of late August to early September in southern England, a medium sized tunnel supporting 1.2 m wide polythene should provide adequate cover. On free-growing soil and to cover plants expected to grow rather strongly, or for a double row, or for two-year-old plants, it may be best to use a larger hoop and 1.3 m wide polythene.

CONSTRUCTION OF HOOPS

Fully galvanised wire of 4 mm (SWG8) or 5 mm (SWG6) should be used, preferably high tensile wire since this makes a more resilient hoop able to withstand pressure and keep a good shape. The wire is best purchased already cut into the required lengths as this avoids the problem of unwinding coils. It is therefore important first to determine the exact size of tunnel and hence the length of wire needed. The length of wire to be inserted in the ground should not be less than 20 cm. On stone-free soils 25 cm can be used, particularly for larger tunnels

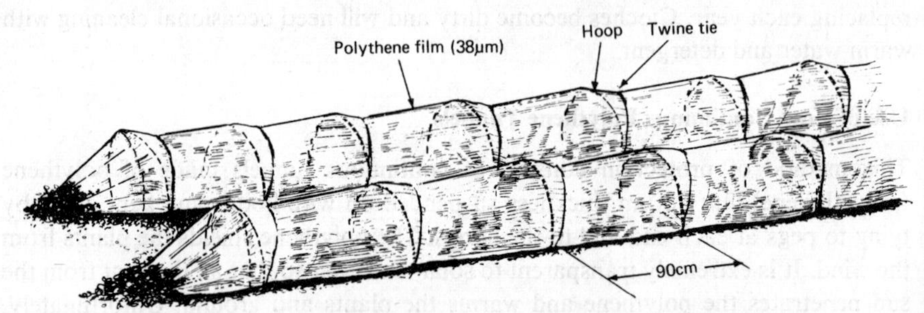

Fig 1 (a) Polyethylene tunnels covering an early strawberry crop

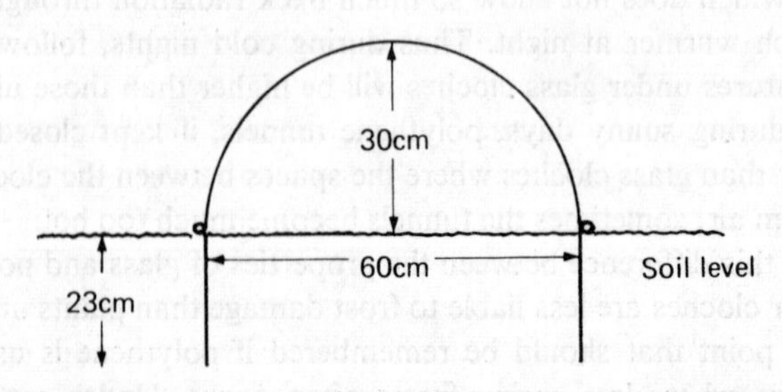

Fig 1 (b) Shape and dimensions for medium sized tunnel hoop

which will then be less inclined to lift under windy conditions. On stony ground it may not be possible to push long lengths of wire into the soil.

A ring on each side of the hoop, facing outwards, is made by holding the correct length for the leg end of the wire in a vice and pulling the wire round a piece of tube 15 mm internal diameter or smaller. If a large number of hoops are being made a small jig can be used as an aid. Making each ring requires about 80 mm of wire but less if a smaller ring is made. The wire can then be bent over a convenient barrel, or an arrangment of nails in a board, to produce the requisite half circle. The curve should be as smooth as possible to avoid damage to the polythene. The actual width and height of the hoops can be readily adjusted when inserting in the soil.

Table 7
Approximate Size of Tunnel Hoops

	Medium (mm)	Large* (mm)
Polythene width	1220	1300
Legs in soil	230	250
Height of tunnel	300	400
Width of tunnel	600	600
Total length of wire	1630	1850
Number of hoops per 25 kg of wire	139	130

*This length of wire is also sufficient to make a flatter hoop about 90 cm wide which can cover a double row of one-year-old plants, if the two rows are planted 30 cm apart.

If a large number of hoops are to be made, it is worth devising a template to hold the wire in the requisite curved loop between two curved pieces on a metal plate. The rings can then be made simultaneously by using a left and right-handed jig to bend the wire around two small iron pegs on the template. Using this type of template the hoops can be made extremely rapidly, whereas using a vice and a single jig, or bare hands, to shape the hooks allows only 20 to 30 hoops to be made per hour, from pre-cut wire.

COVERING THE PLANTS

Autumn herbicides must be used for protected cropping since cultivations are impossible once the hoops are in place. The hoops are put out in straight lines to allow for a tight, compact tunnel anytime during the winter. On large areas where a tractor and planter have been used, a tractor on high clearance row crop wheels is useful to mark out on each side of the plant rows, to clear the hoops after insertion, to run out the polythene and for spraying and fruit cartage. The hoops, tied in bundles of 20 to 30, are placed out at the correct distance over the field ahead of the workers pushing them in the soil in the tractor marks.

The hoops must be positioned not further than 1.2 m apart and, in exposed or windy areas, many growers use them closer to stop wind damage. Two hoops are inserted at the end of each row to take the additional strain at the anchorage point.

The polythene can be placed in position from late January onwards. If a large area has to be covered, a mid-January start may be necessary in order to complete the job by early March. The longer the polythene is out the more likely it is to become weather damaged. But sunny days in February and March can give useful soil temperature increases and coverings before the end of February is recommended if the site is sheltered.

The roll of polythene is secured at the end of each row either by tying to a wooden or iron stake, or by burying a length in the soil. The polythene is then drawn over the hoops, down the row. On a small scale, a hand-drawn trolley which carries the roll of polythene on a freely revolving spindle can be used. The spindle is adjustable for the width of the roll used. The trolley wheels are sufficiently high to clear the wire hoops. The worker pulls the trolley unrolling the polythene ahead of the person tying down. The spindle can be fixed and a good pull taken on the trolley to tension the film which is then eased down over each hoop and secured. For a single handed application the trolley handles can be fitted with two spikes which can be jammed in the soil to hold the trolley in position and the film tensioned.

On a large scale two rolls of polythene are carried on a tractor, which tensions the polythene and two rows are covered at one time. Polythene is sold in rolls of 200 or 600 m. The heavier rolls cannot be easily handled without a tractor or trolley.

The polythene is secured to each hoop using polypropylene binder twine. This should be prepared in the correct lengths required with a simple knotted loop at each end. The twine is slipped over one wire ring, pulled over the polythene, drawing it just below the height of the hoop, and the other end slipped over the other ring. This method is preferable to cutting and tying the twine in the field, a difficult job under cold winter conditions. The correct lengths of twine can most easily be cut by winding the twine over two nails set in a board. The distance apart of the nails must be carefully determined to allow for the two knotted loops at each end. With the tractor mounted machine two continuous rolls of twine are used to criss-cross the tunnels from hoop to hoop.

Polythene film is not usually used a second year; it deteriorates in sunlight and gets dirty. The securing twine can be used again if unhooked and not cut away at the end of the season.

In windy areas there are several methods of preventing lifting of the hoops and damage to the tunnels. Cross ties of twine or lengths of old polythene secured to wooden pegs every sixth row are helpful and can be readily released for passage of spray machinery or pickers, or extra hoops may be used over the outside of the tunnel. In very windy areas more hoops may have to be used, placing them closer in the row. In the south-west, wires are seamed into the edges of the film and secured to the ground by additional pegs. Straw bales can be placed along the rows to act as a low level windbreak.

VENTILATION OF TUNNELS

The polythene can be pushed up between the hoops and ties allowing ready access to the plants for spraying or picking, a big advantage over glass cloches. Because properly erected tunnels are tight to the ground and continuous, very high maximum temperatures may be reached on sunny days in late April and

May, higher than under glass cloches where the warm air escapes through the spaces between the cloches.

Without ventilation the plants become drawn and abnormal in growth, more susceptible to mildew infection and misshapen fruit may result. Many growers endeavour to ventilate the tunnels during the whole blossoming period by pushing the polythene up on one, or both sides on hot days and pulling down again in the early afternoon.

This is laborious and sufficient ventilation may be given by opening short stretches in each tunnel. In some early areas, provided the weather is warm, the tunnels may be partially opened when the plants are in flower and left open during blossoming to aid pollination and fertilisation. It may also be possible to open tunnels on warm days when rain is expected and this will help growth and cropping. It takes about 20 hours to open and close a hectare of tunnels so care must be taken to plan the operation sensibly. The specialist grower with only a small area of tunnels should be able to give adequate attention thus ensuring the best growing conditions for the crop.

It is particularly important to ventilate two-year-old plants, or matted rows, as their growth can become very crowded and etiolated with small and malformed fruit. Experiments using film perforated with slits or holes did not give any advantage over manually ventilated tunnels.

Cultural Treatments for Tunnels and Cloches

In general, heavy cropping varieties such as Cambridge Favourite and Redgauntlet are preferred. Cambridge Vigour, Cambridge Prizewinner and Gorella are earlier but should only be grown where they yield satisfactorily. The usual aphicide spray will be needed two to three weeks earlier than for field plants. *Botrytis* and mildew may not be such a problem on carefully grown plants under protection. Care is needed when applying fungicide sprays to protected plants as the tissues are more delicate. Such soft growth is more liable to leaf and flower scorch than that of field grown plants.

Experience shows that, in most years, the soil often becomes dry during picking, and that irrigation is very beneficial. One of the advantages of tunnels is that they can be easily opened to allow irrigation. If appreciable rainfall occurs during warm days it may be worthwhile opening the tunnels. Irrigation should finish by 1600 hours so that the tunnels can be closed in order to trap sunheat for the night, and to dry out the flowers and foliage if the weather is cool. This is not necessary under warm, dry conditions.

Because of the value of protected strawberries, the plants should be strawed or mulched with black plastic if there is any possibility of soil sticking to the berries. Protected strawberries require careful handling, picking and grading. After picking is finished, if the plants are to be retained to crop the following year, it is important to remove the covers and hoops, clean up the plants, and if necessary, provide irrigation and sprays against red spider or tarsonemid mites, both of which can be troublesome if plants are neglected after picking and allowed to become hot and dry. If Redgauntlet is being retained for an autumn crop, the hoops can be left in position for re-covering to ripen the crop at that time. Cloches should not be left over plants but removed and stacked, to avoid drying out of the plants, particularly important if the plants are to be retained for further cropping.

Film Plastic Single Span Houses

Recently a few growers have produced worthwhile crops of early strawberries, and second autumn crops from Redgauntlet, from plants grown under plastic film houses. The culture of the plants is very similar to that applying to low continuous tunnels, but growing in plastic houses means that picking and all other operations can be done at any time irrespective of weather. Irrigation is, of course, essential and pollination of the flowers may need more attention. The quality of the fruit is generally as good as that obtained from low continuous tunnels. If Redgauntlet is retained during the summer for the autumn crop there are difficulties due to high summer temperatures because the house covers cannot be removed, as can the low tunnels, for ventilation.

It is usual to specialise in the production of strawberries and not to try the crop in rotation with the more usual crops grown under plastic such as tomatoes and lettuce. However, there is no reason why differing rotations should not be tried. The plastic sheet generally lasts two years. The hoops can be readily repositioned and a rotation obtained by moving the house.

Plastic film houses of varying sizes can be purchased ready made and only requiring erection. Or they can be constructed by the grower and full details are given in Leaflet No 17 obtainable free on application to the Lee Valley Experimental Horticulture Station, Hoddesdon, Herts. EN11 9AO. The sheeting usually used is 125μm ultra violet (U.V.) inhibited film, 7.3 m wide and is now sold in 25 and 50 m lengths. Similar size sheets are available in the stronger 150μm (gauge 600) and this sheeting is recommended if the structure is to be covered for two years.

For early strawberry production the house size recommended is 4.3 m wide and 1.8 m high and about 20 m or 40 m long, depending on the length of sheet used. But if it is proposed to use a tractor over the plants then the house must be 5 m wide and made from the 9.2 m sheeting.

The sheeting is supported by 15 mm galvanised steel tubing bent to the appropriate shape and held in the ground by insertion into 25 cm long, 30 mm foundation tubes. The sheet is stretched over the framework and held tight by burying the edges along the two sides of the house and part of the ends. Entrance blinds are constructed to cover the ends and can be opened for ventilation. It has been found that houses up to 20 m long are easier to ventilate and manage than longer structures.

Planting Density and Times

Under most conditions eight rows of strawberry plants evenly spaced across a 4.3 m house appears the best arrangement (Plate VI), with the plants 30 cm apart in the rows. In a 20 m house this gives about 525 plants per house. Some growers may prefer to alter the spacing between four of the rows to allow two slightly wider paths from which the remaining rows can be picked. Various other plant densities have been tested but it is a mistake to crowd the plants, particularly if Redgauntlet is being grown and is to be retained for an autumn crop and as two-year plants. Cambridge Favourite and Cambridge Vigour are also suitable varieties for use in plastic houses.

Plants are generally planted in the open during August to early September and the house covered in mid-February to early March. In the south west it has been found that later planting direct into the house in the autumn and over-wintering under cover has given enhanced plant growth and crop particularly after very wet winters. Local experience will indicate whether later planting is better than earlier planting to give the most suitable size of plant by fruiting time the next spring. It is important to give plenty of ventilation during the early winter to 'chill' the plants and avoid too early regrowth and also to ensure full replenishment of soil moisture before planting. This is particularly important if the strawberry plants follow another summer crop, or are being retained for a second year. A careful check must be kept for aphids and red spider mites which can increase rapidly in numbers and the appropriate autumn and/or spring sprays used.

Plants grown in plastic houses, as under tunnels, need careful attention to ventilation to avoid excessive temperatures. Irrigation is essential and the cheapest method is a simple overhead sprayline suspended from the house framework by wire hooks. Alternatively low level drip irrigation lines may be placed along the rows of plants. If a black polythene mulch is used the lines are laid under the mulch.

Growing in Peat Modules

Many heated glasshouse producers are using modules in which to grow tomatoes, and these used modules are sold at the end of the season. Preliminary trials in the south-west have shown that used peat modules, but not fresh ones (probably because of the high nutrient content), can be used for growing strawberries. One advantage is that the strawberry runners can be inserted in the bags in September to November and provided the bags can be kept watered they need not be moved immediately into the house. Another advantage is that where soil conditions are poor or soil diseases present the modules provide a cheap and satisfactory alternative to growing in the soil. The modules may readily be removed after picking permitting the house to be used for another crop. If Redgauntlet is grown the modules can be stood on polythene outside and kept watered during the summer and brought in for the second crop. This avoids the plants being subject to the very high summer temperatures in the house.

But the use of modules requires knowledge of drip-feed irrigation equipment and techniques, since simple overhead spray lines, suitable for strawberry plants in soil, are not satisfactory for watering modules, but can be retained to aid pollination.

It has been found that sedge peat modules are more reliable than sphagnum ones, not because of the type of peat but because the plastic container is more robust. Eight to twelve plants are inserted in small slits cut in two rows down the side of each module. This positon encourages the berries to hang out and down for ease of picking later on. At planting apply 90 g of superphosphate per bag and water thoroughly. It is most important that the bags should be well watered before and after planting.

The modules are placed end to end in rows down the house, four rows in the 4.2 m, five rows in the 5 m wide plastic house can be accommodated leaving narrow paths for management. It is best to cover the whole floor of the house with black polythene mulch to control weeds and diseases.

Peat modules are difficult to water from overhead spraylines and require drip feed nozzles. They can become soggy if overwatered and extra slits should be cut in the base to help drainage. Strawberry runners root freely in the peat if it is maintained at the right moisture level, but the peat should be examined weekly to ensure that watering and drainage are satisfactory. There may sometimes be problems with high salt concentrations in the peat. There should be sufficient nutrients left from the first crop for root and leaf production until the spring. Although not much information is yet available about feeding strawberry plants it is suggested that when the plants are well grown in the spring daily water requirement will be about 80 per cent of the calculated tomato watering requirement. This figure can be obtained from local ADAS offices. Feed can be added using the $1N{:}2K_2O$ tomato feed diluted at 1:200. To even out water requirements the modules may be laid on capillary matting. There is always some module to module nutrient variation in these secondhand bags following the first crop.

The production of strawberries from vertical socks, polythene tubes full of peat, suspended from framework within houses has also been tried. This permits more plants per unit area. Strawberry plants have also been shown to root well and crop satisfactorily when grown without soil in the nutrient film technique. These methods of production have only been attempted by a few growers and may be more suitable for growers of glasshouse crops who are used to these techniques for other crops.

Pollination and Management

Plenty of warm air movement must be encouraged during flowering to help pollination. Hot dry conditions during blossoming prevent fertilisation and produce malformed berries and overhead damping down may be needed. The yield has been increased when hive bees, or blowflies, have been fed into flowering strawberries in plastic houses in Hampshire. The hive was positioned adjacent to the side of the house, an entrance constructed to allow for the curvature of the plastic, and a small slot also made in the plastic so the bees could fly both in the house and outside.

Blowfly pupae have to be obtained (from fish bait suppliers) ten days before the expected blossoming period. Half should be kept cool at 4°C, the other half, warm at 21°C, spread in sawdust in netted boxes (mesh 4 mm or less) to prevent the blowflies escaping. As the flies hatch the boxes are placed amongst the strawberry plants; the later hatched flies should cover the second half of the flowering period. During the period the bees or flies are fed into the house, no pesticides should be used and the doors should be kept closed and adequate overhead spray irrigation may be given.

During picking, to avoid excessive daytime temperatures, the end blinds can be rolled up and replaced by netting if birds are likely to be troublesome.

If the plants are to be retained in the house for a second year, or for an autumn crop, the leaves should be trimmed off after picking and debris removed, irrigation given and any necessary pest and disease sprays applied to the new growth. It is not always easy to avoid excessive summer temperatures during this period: the shorter houses are easier to keep cool than longer houses. Whitewash shading can be applied to reduce excessive temperatures during the summer and it can be washed off later on.

As with fruit produced under low tunnels extra attention is needed to pick, grade, pack and market carefully the fruit produced from film-plastic houses, in order to obtain a worthwhile premium for this early crop and to cover the extra costs incurred in production.

Late and Perpetual Strawberries

The season for summer fruiting varieties is from mid-May until mid-June for protected plants and mid-June until early August for the main crop. The season can be lengthened by perpetual varieties, ripening during the autumn and from the second crop on Redgauntlet.

Fruit ripe after the main strawberry season has finished never makes as high a price as the early fruit before the main season. But there has always been an interest in the so-called perpetual strawberries. These varieties have the capacity to initiate flower trusses during the warmer, longer days of summer and hence flower and fruit from the normal season until the autumn. In addition to producing flowers continually on the mother plant, they also initiate flowers on the runners so that fruit is formed on these too. Because of the long season, there is never a good weight of fruit to be harvested at any one time and picking costs are high but they are of interest for pick-your-own sites.

The capacity for a long flowering season, or for the production of flushes of flowers in cycles throughout the growing season, is seen in many strawberry varieties and, because of the interest in out-of-season production, plant breeders have been paying more attention to the commercial possibilities of these plants.

Some perpetual varieties have been grown for several years but mainly by amateur gardeners. Various recently introduced English and foreign perpetual varieties are being tested, but none is outstanding, and most are difficult to propagate (page 22).

Autumn Fruit from Redgauntlet

When first introduced, this variety was considered a maincrop, mid-season variety. But if plants are protected in the spring, to give an early crop, successive flowers produced during July and August will give worthwhile crops in September and October, if protected again. Autumn cropping after spring tunnel protection is less certain than after cloching, or plastic film house culture. This ripe autumn fruit of Redgauntlet is produced only in the south and south-west of England, following warm, early springs and under good light conditions. Autumn fruit seldom ripens satisfactorily in the midlands or East Anglia. Cold stored runners of Redgauntlet, if protected, produce plenty of flowers somewhat earlier than from freshly dug runners, and will provide a sequence of fruit from May until November, particularly in plastic houses.

The capacity for second flower initiation in Redgauntlet also occurs on unprotected field plants, but the second fruit generally come so late that it seldom matures except in exceptionally warm autumns.

Sufficient autumn fruit from Redgauntlet is now being supplied to the markets during September until early November for a reasonable price to be maintained. During early August, when the last of the field strawberries are being harvested from late areas, the price is not so good and it appears that a short break between the two crops is desired by the market. Thus the grower for the wholesale market should aim to obtain autumn, or second crop, strawberries from late August onwards.

After the protected crop has been cleared by mid-June, it is essential not to neglect the plants. Cloches and tunnels should be removed and plastic houses well ventilated and watered. The leaves may be mown off and this will destroy any flowers that would produce fruit during August. But unmown plants will crop over a longer period and if strawberries are being sold for pick-your-own, it is better not to defoliate Redgauntlet after fruiting but to allow the plants to continue flowering and cropping during July, August and until the autumn frosts. Irrigation, essential in most years, must be applied to encourage the new growth. A light dressing of nitrogen may be given to help regrowth only on poor soils. If herbicides have been used, the soil should be clean and simazine can be applied again, after irrigation (page 50). If the alleyways are weedy, carefully directed sprays of paraquat can be used to burn off the weed growth prior to simazine application. If straw or plastic mulch has been used, this can be left down to protect the second crop of berries. Runners should be removed, and the plants sprayed against red spider, particularly under plastic film houses.

As the plants come into flower, protective sprays against *Botrytis* are necessary. The plants should be recloched and tunnels replaced, depending on the weather conditions and labour availability, during September.

Experiments have shown that more autumn fruit is obtained from Redgauntlet plants planted early the previous year in August, or July if cold stored, than those planted later. The autumn crop is less than that in the spring, usually about one quarter to one half of that in the spring, provided the plants regrow well and are carefully tended. Redgauntlet plants retained in plastic film houses for the second year often give very satisfactory spring and autumn crops again, provided adequate irrigation is given and pest and disease control is satisfactory and the plants are not crowded. If plants are to be retained for the second year, they should not be defoliated again after the first autumn crop, since this would check the growth of the flower trusses already initiated for the crop the next spring. Redgauntlet autumn fruit is normally of a good size and quality and should be carefully graded. As temperatures fall the later fruit is much smaller.

Perpetual Varieties

There are four named varieties and these have been described on pages 23 and 24. They are all slow to produce runners, so stock is difficult to obtain in quantity and certified virus tested runners are not at present available. It is wise to plant the perpetual varieties away from the main plantations and on sites that are well protected and preferably not liable to bird damage since the fruit ripens over a long period and is difficult to protect both from predators and from grey mould rots.

Culture of these varieties is exactly as for main crop types. Because fruit is carried on the runners the rows may be planted further apart at 1.5 m to allow the runners room to root and be picked without trampling. Some growers remove the early flower trusses: this is said to reduce the mid-season crop and increase yield from August onwards.

Recent small scale trials have shown the benefit to be gained in heavier yields if these varieties are grown on slight ridges and mulched with black polythene. The polythene controls weeds and helps retain moisture during the summer and ridging may help to increase soil temperature as well as improve drainage. The polythene may either be placed over the ridges before planting or set out after planting; small slits are cut for each plant position. The edges of the polythene are held down by pressing into the soil with a disc coulter mulching machine.

The yield of these varieties is often very variable and very dependant both on the season and health status of the plants. Their chief value is on pick-your-own holdings where the time spent in gathering the ripe fruit does not have to be paid for by the grower. For this reason it is not recommended that these varieties should be grown for normal commission sales although they may produce worthwhile returns for the specialist grower with specialised sales outlets.

Harvesting

Harvesting is the most important operation in strawberry production and inefficiency during this very busy time can spoil the whole year's work. It has already been stressed that it is most unwise to grow more area than can be picked. The fruit becomes rapidly over-ripe and once the picking gets behindhand, it is almost impossible to continue, since rotting and over-ripe berries slow down the picking of the remainder of the crop. Promising progress is being made to design mechanical harvesting machines (page 77) and evolve suitable varieties for them. Some fruit needed for processing may be mechanically harvested within the next few years, but as fruit required for the fresh fruit market and canning must be unbruised, it is likely to be several years before hand pickers become completely redundant and then only on the larger holdings where the mechanical harvesters can be used.

Organisation in the Field

Casual labour, mainly women, is necessarily employed for strawberry picking and often the only women available are accompanied by babies, or toddlers below school age. These children can damage the crop and slow down the rate of picking. Where numbers suffice, growers have found it necessary to provide one woman to attend to the children, with a shelter, and a safe play area. In order to get enough pickers, growers may have to employ buses or run their own transport to collect the workers. Care should be taken that insurance is adequate in both cases.

School children may also be employed to pick fruit and, if properly supervised, can be useful when insufficient women are available. Their conditions and hours of work are strictly controlled by County byelaws and a copy of the regulations should be consulted. In some areas, particularly around Wisbech where the crop is sold almost entirely for processing, gang or van labour is used. The grower deals with a gangmaster who is responsible for assembling the gang, providing transport, and supervising the work. The gangmaster agrees a piece work price and conducts all business with the grower.

Whether the fruit is being picked for processing or for dessert it is essential to organise the pickers satisfactorily, provide adequate supervision and to organise a good flow of empty containers and to remove full trays into a suitable shady place, or a cool store as soon as possible. Whether the pickers are on piece or day rates, the central collection point should be planned to avoid unnecessary walking, and to get the fruit into shade as quickly as possible.

Piece Work Harvesting

Pickers need constant supervision not only at the collection point but also throughout the field. The strawberry is so very readily damaged that, even if ungraded fruit is being picked for preserving, all pickers must be shown how to pick the berry by nipping off the stalk and not holding the fruit. The only time the fruit is held is if the crop has been sold plugged, that is the berry is pulled off the stalk in the field.

Harvesting fruit for preserving or canning is normally done by piece work and the berries are usually picked direct into trays, or plastic baskets, holding 1–5 kg of fruit; very small or damaged berries should be discarded into the alleyway. The pickers are paid as they bring their full containers to the collection point where the unit is weighed and inspected. These piece work rates are based on a picking rate of about 6–10 kg per hour, with higher rates for plugged fruit picked off the stalk. The rate depends greatly on the weight of the crop and condition, berry size and availability of pickers each year.

Ideally, the strawberry fields should be picked over daily in really hot weather, and two or three times weekly under cooler conditions. In practice, picking is often done only once per week, particularly for processing, but it is most unwise to grow more strawberries than can be harvested with the pickers likely to be available. At least 20 to 30 pickers would be needed per hectare for picking and grading an average crop into punnets; more workers would be required if more careful grading and packing was to be done. Rates for picking into punnets and grading the fruit average about 4–7 kg per hour, but there can be great individual variation.

Varieties vary greatly in the speed with which they become over-ripe. The popularity of Cambridge Favourite is due in no small measure to the way it will stay on the plant when fully coloured for several days, without becoming too soft or dark.

Fruit for processing should be picked firm ripe with one to two-thirds of the surface showing colour. All blemished and rotted berries should be thrown down in the alleyways.

Harvesting Dessert Fruit

The majority of growers pick the fruit directly into the punnets already arranged in the market tray. This means that little or no grading of berries can be done; the pickers generally only leave very small, damaged, or rotted berries on the plant, or, if picked, drop this fruit in the alleyways. Although this method results in the maximum output per picker, it has the disadvantages that the tray may get dirty or wet, pickers tend to do a final 'top up' of the punnets, using the bigger and more attractive berries, and often put damaged fruit into the punnets. If pickers are being paid piece work, there is always a tendency to skimp the grading, hold the berries instead of the stalk and to pack them into punnets as quickly as possible. The only quality control that can be exercised under these conditions is at the collection point, where the trays should be examined as they are brought in by each picker. At the beginning and occasionally thereafter, in addition to a surface inspection, a punnet should be removed and the contents examined for bruising, retention of damaged fruit and uneven grading.

It is essential to exercise as strict a control as possible over the pickers, and those persistently bruising berries, or packing unsuitable berries, should be dismissed. The biggest problem is to stop pickers holding the berries.

To some extent the crop sets the standard of picking and packing, since well grown, not too leafy plants, with readily seen, large berries are obviously easier to pick and grade than weedy, crowded plants with small or diseased berries.

Many growers prefer to work on a day-rate system with a bonus for each tray properly picked and presented. This method may well be the only way to get punnet fruit carefully picked and graded, but it involves some ingenuity to arrive at a system that encourages rapid work at reasonable rates.

The punnets should be placed into the market trays ahead of the pickers. Where there is plenty of storage space the market trays can be prepared before the picking season, or during wet weather, and stacked ready for the pickers. Even if this is not done it is best to have one woman preparing the trays—a useful job for an incompetent picker, or an older woman—rather than let all the pickers do this.

Laying the fruit into the punnet requires experience. The berries are laid in the corners first and then when the bottom has been covered, the next layer is placed in position. The berries should be laid in with the stalk slightly downwards, so that the final appearance is one of the attractive colour of the berries with the minimum of green stalk and calyx showing. Some pickers try to pack the top layer of berries stalk downwards. Although a most attractive appearance results, every berry has to be handled to be placed in position and must therefore be bruised.

Grading Dessert Fruit

Growers producing out of season and quality crops, generally endeavour to grade the fruit properly. Some form of grading may also be attempted by growers of main crop strawberries, who are concerned to pack to a better standard than the average.

Grading may be done in the field by the pickers, or in a packing shed. The latter is uncommon but is practised by some specialist producers of early and autumn fruit. Each picker is provided with carriers which are generally wooden trays fitted with a handle. In the bottom of each tray is placed a folded cloth. The pickers are instructed to pick every berry that is two-thirds or more coloured even if it is damaged and to place it carefully in the tray in a single layer. As the trays are filled they are stacked and taken to the packing shed. This need not be elaborate but should be well shaded, dust free and airy. In the shed, tables are needed at a suitable working height. The more deft and experienced pickers are kept in the shed to repack the fruit. They have the market trays complete with punnets in front of them and rapidly transfer the berries to the appropriate grade, checking each fruit for damage and downgrading as necessary.

Although this method involves double handling of each berry it can ensure excellent standardised grades and all the damaged fruits are removed from the field. It is particularly valuable where labour is coming and going at all hours, as it enables those in the field to concentrate only on removal of fruit and not to need instruction in, and supervision of, grading. Each berry should be examined for soundness; even the smallest infection by *Botrytis* will result in a wholly rotted berry by the morning and other berries in the punnet may become infected. Berries marked or damaged by boxes, or feet, slugs, beetles or birds must also be ruthlessly downgraded and discarded, not mixed with sound fruit. This is particularly important during wet weather when rots develop rapidly.

Tray Carriers

Although many growers allow pickers to use the cardboard market tray containing the punnets in the field this is often rather large to fit between the rows of plants so that fruit gets damaged. The tray may also get wet or muddy and is somewhat flimsy for field conditions. Some growers use tray carriers made of wood, iron or aluminium alloy construction to hold the tray. These carriers can best be made to go right over, and clear of, the row of plants, or to fit in the alleyways (Fig. 2). They must be of sturdy construction as pickers tend to lean on the handle when moving forward, or standing up. Some narrow carriers are designed to hold only punnets which are then repacked in the market containers.

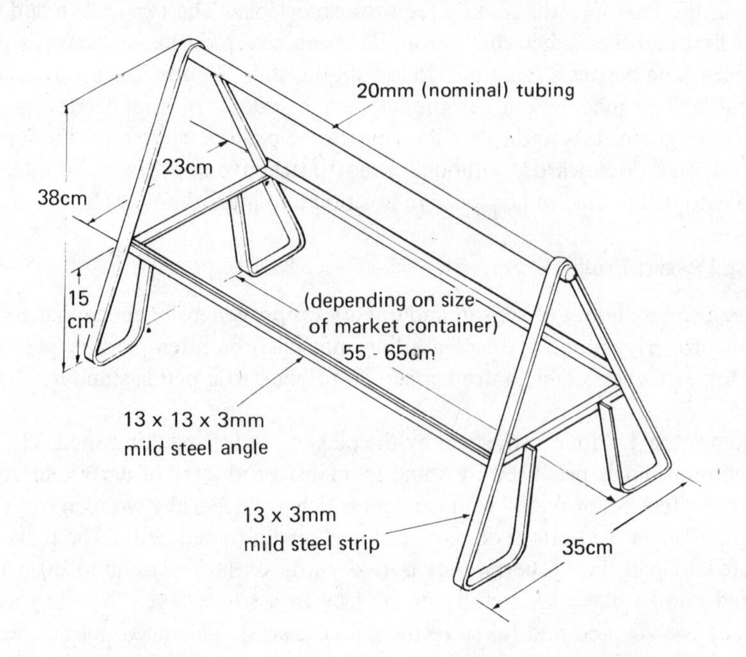

Fig 2 Tray carrier for use over plant row when picking direct into punnets in market trays. Net weight of tray carrier about 2 kg.

Punnets

These are available in plastic, moulded paper and waxed cardboard usually to hold 100–125 g, about 250 g and 500 g of fruit. Plastic punnets are the most popular being impervious to moisture but it is important to select a good design with some springiness so they will pack into the market tray without movement. Waxed cardboard punnets sometimes damage the berries with their edges as the fruit is placed within the punnet and moulded paper ones are absorbent and not always well enough ventilated.

Market Trays

These are of fibreboard or wood, holding a single layer of punnets. It is best to use trays to fit the size of punnet, or punnets, to be used. Not all trays will contain satisfactorily both square 500 g and rectangular 250 g punnets. The corner post stacking tray is best as this permits free passage of air over the fruit and also prevents the tray above resting on fruit below, particularly if this is piled rather high in the punnets. Fibreboard containers should not be stacked too high or the bottom ones maybe crushed. Most market trays now used for strawberries are non-returnable and often are covered with a see-through window cover to stop pilfering and the berries moving about during transport.

EEC Standards for Fresh Strawberries

These statutory regulations apply to strawberries supplied fresh to the consumer through normal wholesale and retail trade. They do not apply to fruit for commercial processing, or to sales by growers on their holdings to consumers for their personal use. The regulations specify minimum requirements for three classes, Extra, Class I and III (no Class II) by quality and size requirements. Full details are in the appropriate leaflet obtainable from the Horticultural Marketing Inspectorate of MAFF, or from Divisional Offices. Inspectors and HAOs of ADAS are available to help with advice on grading including the labelling requirements. The inspectorate have powers to examine fruit offered for sale, both in wholesale or other markets and in retail outlets, and to ensure that it meets the minimum statutory requirements in force.

Mechanical Harvesting

Many growers are now interested in the developing of mechanical harvesting to avoid problems in gathering, organising and supervising large numbers of pickers, and testing in the field of prototype machines is being carried out in the main strawberry growing areas by ADAS and commercial firms.

There are several well defined stages in mechanical harvesting:

1. Removal of the whole truss of fruits from the plant.
2. Separation of the fruit from the leaves and trash.
3. Singling of the truss into individual berries with about 5 cm of pedicel remaining.
4. Removal of the calyx (capping).
5. Sorting of the capped berries for various outlets.

A prototype harvester was designed at the National Institute of Agricultural Engineering and comprised a cutter bar to cut off the plant and an elevator to carry the plant onto a cross belt where the leaves and trash were separated from the fruit trusses by a current of air (Plate IV). The separated trusses were then collected in trays. A commercial firm is now developing and improving the whole system in conjunction with ADAS and growers in England and Scotland.

Design of the singulator has proved difficult but is now promising success, and is fitted on the harvester to deliver from the field strawberries still on their stalks. The trays are then taken to a capping-line in a building where they are fed onto a conveyor, through a washer and then onto the capping machine. The present decapper traps the pedicels of the singled fruit between contra-rotating rollers and carries them to a band knife, which slices the berry just below the calyx. At the end of the line there is a collection of green, white, pink, red and over-ripe fruit which has to be sorted and graded into the requirements of the various outlets. The sample is then virtually ready for the processing line, without further sorting or wastage. The labour requirement and general economic level of the whole operation have still to be assessed.

In the field the matted row of strawberries must be restricted to a clearly defined band of plants whose crowns can be accommodated within the width of the cutter bar to allow the trusses to be lifted and cut. The ground needs to be level to assist the even pick-up of the fruit and prevent the cutter bar digging into the soil. Ideally there should be no loose soil, stones, or clods to be picked up, elevated onto the machine and cause contamination of the fruit. A good complete herbicide programme is essential. Weed control has to be effective as the presence of weeds upsets both the elevation of the trusses and their separation from the foliage. Very good control of fruit rot is necessary as there will be susceptible, fully ripe fruit on the plant before harvesting. Experiments have shown that the cut-over plants produce satisfactorily in following seasons.

The strawberry variety to suit the present mechanical harvesting system best would be one with a short ripening season, an upright truss bearing relatively few fruits with pedicels (stalks) over 5 cm long, prominent shoulders and firm flesh. In the course of the next few years mechanical harvesting of strawberries seems certain to progress, particularly in those areas where the crop is grown mainly for processing outlets.

Marketing

There are three main channels of distribution to the consumer of fresh fruit: through wholesale channels, or direct to the public by PYO or farm shop sales, or direct sales to shops and supermarkets. Strawberries for retail sale are consigned to wholesalers for sale on commission but because of the perishable nature of the produce, direct sales to local shops, local wholesalers, or to the public, can be beneficial to the grower and provide the consumer with a fresher product.

Wholesale Marketing

Strawberries for the fresh fruit market are usually picked and packed direct into non-returnable containers called punnets or chips, which are transported to the markets in crates, trays or carriers. These outer containers are usually non-returnable. Quick transport to the market is important, preferably overnight in unsheeted loads so cool night air flows over the fruit. Wholesalers specialising in strawberries generally arrange transport from the main areas so that strawberries are handled separately from other produce that is not so perishable. The costs of selling through wholesale markets will include the salesman's commission, usually 10 to 12½ per cent of the gross selling price, and a market toll per packet, which can vary according to the size of the container. A porterage charge is frequently made even if the produce is collected by the purchaser. The grower also has to pay transport costs to the market, or where the wholesale salesmen arrange collection, there may be a collection charge. Invoices from the salesmen should show clearly the gross selling prices of the various packages of fruit and the charges which have been deducted. In some instances the salesman will provide returnable trays and covers, possibly stamped with his own brand name, the cost of which is sometimes included in the commission, or for negotiated charges.

Salesmen tend to specialise in the products they handle and it is important for the grower to keep in touch with the requirements of the market and to choose salesmen and market centres most suitable for his transport arrangments and the type of fruit produced. When possible, the grower should visit the salesman during the marketing of his crop to assess how his fruit has travelled and to compare grading, quality and pack in relation to other market consignments. Growers should also encourage their salesman to notify market requirements to them, and, to comment on grading and quality of fruit as received.

Although punnets are commonly referred to in terms of the weight of fruit they hold, there is no statutory requirements for a grower to sell by weight when selling through wholesale channels. It is in fact a common practice, which is accepted in the markets, for wholesale sales to be by the punnet–by volume—without the contents being weighed. However, if the fruit is sold by weight and marked accordingly, then the contents must be made up to the correct weight for the particular size of punnet, and must correspond with any marking of the weight. Under the Weights & Measures Act 1963 all retail sales (including direct sales by a grower to a retail customer) must be by weight, and the customer must be informed of the weight before he pays for, or takes possession of, the fruit. The way in which this may be done depends upon the weight of the empty punnet. Where this does not exceed the amount shown in the table, the fruit may be weighed in the punnet, and the gross weight marked on a ticket or on the punnet, or a special weighing machine may be made available for the customer's own use. Where the weight of the empty punnet exceeds the amount listed, the net weight of the contents must be marked on the punnet.

Gross weight of punnet and fruit Kg of gross weight (g)	Maximum container weight per (g)
up to 250	120
250–1000	100
1–3000	90
Exceeding 3000	60

Pick Your Own (PYO) and Farm Shop Sales

Over the last few years there has been a large increase in the area of strawberries grown to be harvested entirely by the customers. Many of these plantations are on arable or general farms with no other fruit crop, but well situated relative to a good road system, or close to urban areas. Provided the soil is suitable and pre-planting preparation is thorough and that certified runners are used the crop is not difficult to grow using machinery available on the general farm. A sprayer may need to be purchased. The real difficulty in strawberry production has always been the harvesting and if the public can be enticed to pick-their-own then this problem is solved.

Organising a PYO field of strawberries requires planning and attention to detail. Assuming that several fields of suitable soil are available access is the most important consideration. Most customers come in cars and need safe approach lanes and gateways, clearly posted and organised car park, or parks, near to the crop. Although strawberry beds can be retained longer for PYO than market use, since slightly smaller fruit is of less importance, most beds will need replanting every four to six years, so a rotation of cropping fields and suitable car parks must be planned from the start. A ratio of 1:15 car park to cropping area will only just suffice on busy, sunny weekends. Access from the public highway must be carefully planned and advice sought from the County or District Council Planning Department. It is particularly important that PYO holdings do not create traffic hazards. Planning regulations limit the size and situation of placards advertising the fruit; again information is available from the local Planning Department.

Good clear notices and directional signs are all important, but an annual charge must be allowed for advertising the situation of the holding, the dates and times of opening and prices. Many growers find that these costs are similar to commission costs for wholesale selling on the crop. Local newspapers are widely used, but growers of larger areas and with other PYO crops frequently buy local radio and TV time to advertise the holding. Coach companies can be persuaded to run strawberry trips and WI and Townswomen's Guilds may organise outings to pick fruit. A robot phone in the farm office, or better, office staff to answer enquiries and give the daily prices and availability of fruit is essential.

On the larger holdings it is essential to have efficient pay points and enough interested staff on duty to avoid delay at busy times. Many growers provide a free mushroom basket; this holds up to 2 kg of fruit, is easy to use and is of constant weight for pay out checking. Four baskets can readily be carried by one customer if a large amount of fruit is needed for preserving or freezing. The Weights & Measures Act (page 80) applies to sales of strawberries and the appropriate officer of the local Council can advise as to its application. Weighing machines must meet legal requirements and may be inspected from time to time.

Pricing of PYO crops can be difficult; the price must be high enough to cover profitably all the overheads as well as the direct growing costs and yet must ensure that the crop is kept well picked. The wholesale market prices are a useful guide.

Another difficulty is the adjustment of advertising and attraction of customers to the ripening of the crop. Most PYO farms only grow Cambridge Favourite because of its high yield, strong plant constitution and ability to remain ripe for several days, but there is much merit in having about one tenth of the area planted to Cambridge Vigour or Gorella, both earlier to ripen than Cambridge Favourite. This enables the public to be invited to the holdings about a week before Cambridge Favourite is ripe. Many customers are brought along by word of mouth information and these early crops ensure more pickers by the time Cambridge Favourite is fully ripe. Because the picking rate does not matter to the grower, the smaller fruit on older plants of Cambridge Vigour are not a disadvantage and its better flavour, and the red flesh and flavour of Gorella may be good selling points.

The perpetual varieties are worth trying for PYO and it may also be worthwhile having an area of a suitable early variety covered with low tunnels to start the season. Domanil and Redgauntlet are slightly later than Cambridge Favourite and small areas may be worth growing to extend the season.

The smaller and more specialist grower might offer strawberries grown especially for their flavour, or only with organic manures (FYM) to attract special customers. PYO selling requires organisation and long hours of work for six days a week—many PYO holdings are shut on Mondays if the crop has been cleared over the week-end—for several weeks.

Selling the crop through PYO is never certain. Near urban areas the biggest demand is at the week-ends, although evenings are also popular. Bad week-end weather can result in much spoilt fruit and difficulties the following week. The best PYO farm is one with a range of horticultural crops, well situated relative to a retired, as well as urban, population. The retired people like to pick mid-week. Some growers use both PYO and their own pickers to market the fruit. A proportion of the fruit can also be displayed in baskets or punnets and offered for sale at the pay out point. Self pickers often get carried away and pick more fruit than they wish to buy. This can be put on the farmshop sale after checking, or if of a suitable grade added to the market load.

Many PYO holdings start with strawberries but then diversify with other soft fruit and vegetable crops. ADAS publish a useful booklet *Direct Farm Sales to the Public* that can be obtained through Regional and Divisional Offices and which provides information on the many regulations relating to planning, weights and measures and other Acts that have to be complied with if this form of selling is adopted.

Direct Sales to Shops, Hotels and Supermarkets

If the grower can negotiate suitable long term agreements these methods are valuable as commission and porterage are eliminated. For supermarkets it may involve special grading and punnet over-wrapping requirements, fruit cooling, and delivery costs. If the extra specification and work interfere with supervision of the picking and packing the grower may not be able to maintain the agreed grading standards. Growers who are able to cool their fruit, by stacking the trays on pallets and loading into cold store as soon as possible after picking, should be able to supply those supermarkets who wish to sell these better quality strawberries from a cool counter. The benefits of cooling strawberries after picking are lost if the fruit is subsequently handled through markets, or shops, without cool conditions; the berries then sweat and look unattractive.

Cooling and Storage

When strawberries are picked on a normal day in July the field heat of the berry flesh can be as much as 37°C. To reduce the flesh temperature berries should, ideally, be picked early in the morning when their temperature is low and immediately placed in cool shade, or a cooled store.

In practice much of the fruit has to be picked during the heat of the day. This fruit should be moved as soon as possible into cool shade. Even a simple suncover on the collection trailer will be worthwhile. On small holdings it may only be possible to stack the fruit in a cool shed but some extra roof insulation would provide cooler conditions.

If the fruit is going to be handled through a cool chain right to the point of sale the aim should be to extract the field heat as quickly as possible to get the fruit flesh temperature down. The fruit should be taken to a store as quickly as possible to be cooled down to 4°C within six hours of picking. During this process about one per cent of the fruit weight is lost; slower cooling increases weight loss.

The rate of cooling will depend on the original temperature of the fruit and the equipment available. In many cases these will be cold stores normally used for top fruit storage. The rate of cooling depends greatly on the flow of air over the fruit; pallets should therefore not be stacked tightly. Corner piece trays to hold the punnets are more suitable than tight stacking fibreboard trays. A guide to maximum loading rates of fruit per hour can be calculated by multiplying the store volume (m^3) by the appropriate factor from Table 8.

In practice most top fruit stores are likely to have fan flow rates sufficient to achieve the necessary cooling, for coping with the weight of fruit picked and speed of loading on most farms. The cost of running the store for cooling strawberries is

Table 8

Maximum loading rates (kg fruit/m³ h) into a cold store

Ambient Conditions	Air Temperature °C	Cooling Period 3 hours	6 hours
Hot	21	2.65	2.80
Warm	16	4.00	4.22
Cool	10	8.00	8.40

minimal. The use of a down hanging polythene curtain to restrict the flow of cold air from the cooler to the immediate vicinity of the pallets is worthwhile. Care should be taken to open the store doors as little as possible when bringing in the fruit. One tonne of fruit packed in wooden trays or pallets requires about 8 m³ space.

Strawberries of good quality, picked firm ripe under good conditions and promptly cooled to 4°C as described can be safely held in store for a few days. Up to five days can lapse between picking and consumption if the fruit is cooled, stored, transported in refrigerated transport and put on sale in a cool counter. This storage is particularly helpful to meet weekend peak shopping and to help avoid gluts or wet weather problems. Fruit has been stored for much longer than this but the final quality is not satisfactory. Fruit held in store and then marketed through normal unrefrigerated transport and shops deteriorates very rapidly. Also it is most unwise to try to store fruit that is fully ripe, or overripe, or of poor quality.

Fruit to be sold through non-refrigerated shops and markets or from the farmshop however, can benefit if some of the field heat is removed after picking. In this case the aim should be to cool the fruit to 10°C, so as to avoid too much sweating when the fruit is out of the store. This may be done in top fruit stores; alternatively a refrigerated lorry could be hired and placed at the farm in a convenient position. Some growers have purchased such vehicles secondhand. The aim should be to cool the fruit, not with a view to holding it in store, but in order to improve the quality and reduce the rate of deterioration. Advice on use of cold storage for strawberries can be obtained from Mechanisation Advisory Officers at MAFF Divisional Offices. Fuller technical details are given in the publication *Refrigerated Storage of Fruit and Vegetables.*

Fruit for Processing

Usually buying agents contract with growers to purchase part, or all, of the crop on behalf of the factories. The agent arranges the delivery of empty trays, or tubs, throughout the picking season and the transport of the fruit to the factory each night and often during the day. This has the advantage that the grower can concentrate on getting the fruit picked to the required standard; it is the agent's job to collect up the fruit and cope with the empties. Some growers of large areas of strawberries deal with the factories direct. Sometimes the grower will take two picks of fruit and send these to the wholesale market and then the remaining picks to the factories. Canning and freezing factories do not always want the very large berries, particularly of Redgauntlet, that often come in the primary fruits. Fruit can be put into tubs, to which sulphur dioxide solution is added as the preservative, on the farm. This fruit is used for some grades of jam.

Pest and Disease Control

In order to start a plantation as free from pests and, particularly, diseases as is possible, throughout this book the importance of only planting certified runners has been stressed. New plantations should be sited as far as possible away from existing plantations if they are harbouring pests and diseases. Land that has never grown strawberries before is unlikely to harbour the soil disease red core, but unfortunately the strawberry can be attacked by several pests and diseases that also attack other crop plants or survive on weeds.

Well managed plants, in well drained soils are less likely to be troubled by many pests and diseases but there is one disease Botrytis fruit rot that is ubiquitous and readily attacks flowers and fruit particularly under wet conditions; protective control measures should be routine each year against this problem or much of the crop may be unsaleable with fruit rots. Many of the other pests and diseases need only be controlled if they are present and likely to become a problem.

Unnecessary spraying of pesticides should be avoided, particularly those that kill beneficial insects or bees.

HMSO, for MAFF, publishes early each year a useful book—Agricultural Chemicals Approval Scheme List of *Approved Products for Farmers and Growers** giving details of proprietary chemicals which have been approved for specific purposes and these pesticides should always be chosen. All approved chemicals carry on their labels this mark. The manufacturers' recommendations about use, mixing, application and safety measures must be strictly followed. To ensure that the harvested strawberries do not contain harmful pesticide residues it is important to follow the directions on the strength and frequency of application, and to observe the minimum interval between the last application and anticipated harvest date.

All chemicals should be used with caution and stored in a safe locked place. Empty containers should be carefully washed out, beaten flat if necessary, and disposed of safely. Sprays should not be allowed to contaminate ponds or

*Appendix I

waterways. Spray operators should wash off any concentrate that may fall on their skin. THERE ARE STATUTORY OBLIGATIONS AFFECTING EMPLOYERS AND EMPLOYEES WHO USE CERTAIN POISONOUS SUBSTANCES AND THESE SPECIAL REGULATIONS MUST BE CAREFULLY OBSERVED. A leaflet *'The Safe Use of Poisonous Chemicals on the Farm*'* is available. Growers should consult their local ADAS Horticultural Adviser if in doubt about the identification of pests or diseases and their control. Routine checking and careful observations, together with speedy and adequate control measures should help to prevent serious, or persistent, problems occurring.

Pest and Disease Identification

The accurate identification of pests or diseases is essential if the correct control measures are to be taken, not only to obtain an adequate control but also to avoid the unnecessary use of pesticides. As many materials cannot be used near or during the picking season, preventive or prompt remedial action is often necessary. Many pests and diseases give similar symptoms and must be identified accurately so the correct control measures can be applied. MAFF publish many advisory leaflets describing and illustrating pests and diseases which attack strawberries (Appendix I).
These leaflets are more frequently updated than this book.

A large range of pesticides is available for the control of strawberry pests and diseases but as they are altered and added to over the years they have not been included in the following descriptions of pest and disease damage and control. ADAS publish a booklet *'Fruit Growers Guide to the Use of Chemical Sprays'* which is updated frequently and available through Divisional Offices. The grower should consult this publication together with the annually published book *Approved Products for Farmers and Growers,* obtainable from HMSO agents, in order to select a suitable pesticide for the control of any particular pest or disease.

Pests of Major Importance

Aphids

Several different species of *aphis,* often called greenfly may occur on strawberry. Only two are of economic importance. These are the strawberry aphid (*Chaetosiphon fragaefolii*) and the shallot aphid (*Myzus ascalonicus*).

The strawberry aphid is the main vector (carrier) of yellow edge and crinkle virus diseases. It causes no obvious feeding injury and is mostly seen on the young, unfolded leaves when it is just visible to the unaided eye. It is a creamy white aphid characterised by the many knobbed hairs projecting from the head and back, easily seen if the aphid is examined from the side using a magnifying

*Appendix II

glass. It may be found on plants all through the year; peak populations occur in early summer on fruiting beds and in September on first year plants.

The shallot aphid is pale greenish-brown and feeds on the young leaves in the crown. It breeds on the plants throughout the winter and spring, migrating to other host plants in late May. This aphid may cause severe damage after mild winters when high populations produce extensive, early leaf distortion and stunting of the plants; the flowers are malformed and little fruit develops. This damage resembles, and may be confused with, crinkle virus disease symptoms.

CONTROL

Control of strawberry aphid is important in order to reduce the spread of virus diseases. The pesticides used will also deal with any other species of aphid which may be present. Outdoor fruiting beds should be sprayed in spring, after some growth has taken place but before flowering, using a suitable aphicide. Alternatively certain granular pesticides can be applied to the soil provided an interval of at least six weeks can be left before picking. On older beds an autumn spray may be worthwhile if aphid populations have built up during the season; this spray might reduce the need to apply control measures the following spring, a worthwhile advantage where the crop is grown under cloches or tunnels.

If earlier control measures have not been applied and a severe aphid infestation develops during fruiting, causing plant damage and sticky fruit, control can be obtained with one of the pesticides of short persistence (i.e. one requiring an interval of only a few days between application and harvest). Runner beds must be adequately protected from migrant aphids which might be carrying virus. At least three sprays are required in mid to late May, mid-June and early July using one of the more persistent aphicides recommended as sprays for fruiting beds, or use of granular aphicides.

Red Spider Mite

The species of red spider mite found on strawberries is *Tetranychus urticae*. It overwinters in the adult female stage sheltering in cracks in the soil, on the undersides of the lower leaves, or on the straw and weeds. These brick-red coloured mites become active in April or earlier on cloched crops. The greenish summer form of the mite and their dull white eggs, can easily be seen with a magnifying glass on the underside of the leaves where the active stages feed, causing extensive, patchy mottled areas. Up to seven overlapping generations may occur during the season. Some cultivars, such as Cambridge Favourite, are very susceptible and in some seasons, in the absence of control measures, protected crops under cloches or tunnels may become heavily infested before picking. The growth becomes poor and stunted, the whole plant appears bronzed and sometimes becomes covered with webbing. In hot and dry seasons severe damage may also occur on outdoor plants. (Plate VIII). Advisory Leaflet 226 also describes red spider mite and its control.

CONTROL

The organophosphorus pesticides recommended for aphid control in spring may also control red spider mite on some holdings. However, mite populations resistant to many of these pesticides and also to certain acaricides, are now widely distributed in the main strawberry-growing areas. The choice of pesticide therefore depends on the strain of mite which is present. Some pesticides only give good results when used in warm conditions (15°C or over). Some pesticides may reduce fruit set if applied to open blossoms, particularly on protected crops, or cause slight leaf scorch.

On crops in the open, the acaricide may be added to the aphid spray applied before flowering (check compatibility of the mixture). In some seasons a second spray may be required after harvest and this will reduce the risk of serious infestations in the following spring. With crops grown under cloches or tunnels where conditions favour a rapid build up of the mite, two pre-blossom sprays may be required as a routine.

Where an autumn crop is to be taken on Redgauntlet a suitable acaricide should be applied when the new growth appears after mowing off. In hot, dry weather a further spray may be necessary before fruiting. On runner beds the first spray should be applied in May preferably followed by a second application two to three weeks later. A further spray may be required in some seasons.

Less Frequent Pests

Strawberry or Tarsonemid Mite

This microscopic mite (*Tarsonemus fragariae*) is troublesome in some areas, particularly in warm summers; it feeds in the young folded leaves, causing roughened, rust-coloured areas and a characteristic downward leaf-edge curl on some cultivars. Later in the season the crown leaves may turn brown and die and the plant may lose its fresh green colour, the older leaves being puckered and wrinkled. With severe infestations the whole plant may become stunted. Damage is unlikely to be obvious until after fruiting of the first crop. Cloched plants and older beds are particularly liable to attack. When heavily infested, Redgauntlet may produce little if any fruit in the autumn crop (see also AL 584).

CONTROL

Sprays should be applied thoroughly after picking using drenching, high volume applications of the approved pesticide. Where an autumn crop is to be taken, care must be taken to follow the approved interval between spraying and mowing, or harvest.

Nematodes or Eelworms

Nematode damage to the leaves and crown of the plants may be confused with shallot aphid damage but can be distinguished by the absence of aphids and cast skins. Nematodes are very small and cannot usually be seen with the unaided eye, nor easily with a magnifying glass. Advisory Leaflet 414 *Eelworms on Strawberry* also gives details.

Leaf nematodes

These species (*Aphelenchoides fragariae* and *A. ritzemabosi*) feed on the young developing leaves, causing damage most noticeable in the spring when growth is slow. Affected leaves may be puckered, distorted and fail to open normally, some leaves may show greyish or silvery feeding areas on the upper surface towards the base of the midrib. The main crown often becomes blind and many secondary crowns then develop, forming an open-centred plant which crops poorly. Runners from infested plants will also be infested but may look healthy. When this eelworm occurs together with the bacterium *Corynebacterium fascians,* cauliflower symptom is produced. Instead of developing normally the growing point produces growth rather like the curd of a cauliflower.

Stem nematode (*Ditylenchus dipsaci*)

This nematode generally causes a marked corrugation of the leaves and a shortening or thickening of the leaf stalks. There are several races of stem nematode, differing mainly in the host plants on which they live. It is known that the oat, red clover and narcissus races can infest strawberries. Parsnips, rhubarb, broad and french beans, onions, vetches and mangolds are also important hosts of this nematode, as are many weeds. Strawberries should on no account be grown on land where any of these crops have previously been infested by stem nematode. Very severe attacks can occur in fields where strawberries are replanted annually. Runners should never be taken from infested plants. (Plate VIII)

CONTROL

To avoid nematode problems it is essential that only clean runners are planted as little can be done to clean up infested plants. If only small numbers of plants are affected in fruiting beds, these can be dug out and burnt to reduce the risk of spread. There are no satisfactory control measures for heavily infested beds. For light attacks a granular pesticide is available which may be applied within seven to ten days of planting the runners and incorporated into the soil by overhead irrigation immediately after application, as directed by the manufacturer. In a season of above average rainfall a second application about six weeks later may be necessary. It will not control nematodes in older plants when they are severely infested nor in situations where the soil is very heavily infested.

Chemical soil fumigation may also reduce nematode numbers in the soil. Plants should be lifted as soon as possible after cropping and the soil injected in the following late spring or summer. Better results can be expected if the soil is kept free of strawberries for at least one year.

Free-living soil nematode

The dagger nematode (*Xiphinema diversicaudatum,*) is a soil-inhabiting, root-feeding species and locally important because it transmits arabis mosaic virus. The nematodes do not themselves seriously damage the plants unless unusually large numbers are present, but if they are carrying the virus they will infect the plants causing severe crop loss. Symptoms usually occur in well defined patches. *X. diversicaudatum* can also transmit strawberry latent ringspot virus.

CONTROL

The dagger nematode can be controlled by soil-fumigation carried out at least six weeks but not more than six months before planting. Autumn application has been shown to give good control of the nematode. These fumigants will kill other nematodes and other soil pests, weed seeds and plants and thus cannot be used after the crop is planted. They should be injected at a depth of 15–20 cm into well worked moist soil where they vaporise, killing the nematodes. Soil fumigation can never completely eliminate the pest because nematodes in the deeper layers can re-infest the treated soil. After treatment the surface should be rolled or sealed by thorough watering. Three weeks after injection the soil must be thoroughly cultivated, followed by a second cultivation two weeks later. The soil must be completely free of vapour before strawberries are planted. Planting six weeks after injection should be safe provided that the soil has been well cultivated and no trace of the chemical remains.

Soil fumigants are corrosive and strongly irritating to skin, nose and mouth and care must be taken when handling them and while cleaning machinery after use. A hand injector can be used for small areas; the single spike is pressed into the ground at 30 cm centres in staggered rows, the plunger then delivers a pre-set amount of liquid at each point. A tractor-mounted machine can be used for large areas. The chemical is gravity fed from a tank through polythene tubes to hollow injector tines.

Tortrix Caterpillars

A number of species occur but the strawberry tortrix moth (*Acleris comariana*) is the most commonly found. The caterpillars feed on the leaves, webbing them together. Smaller numbers may also attack the flowers. They emerge in April or early May from eggs which have over-wintered on the plants. The very active young caterpillars are white with black heads later becoming green with a shiny yellow head. The adult moth emerges during late June and July and eggs are laid mainly on the undersides of the mature leaves. A second brood of caterpillars may be found on the plants in August and September. This pest is particularly numerous in strawberry-growing areas near Wisbech. Other species including *Clepsis spectrana* and *Olethreutes lacunana,* are found in some areas and appear to have increased in importance in recent years. These overwinter as young, dark brown caterpillars causing damage very early in the season, particularly on protected crops. Caterpillars of these species are frequently found damaging the flowers.

CONTROL

Where tortrix is a problem spray in April or when damage is first seen before flowering. More satisfactory control will be obtained with most pesticides if they are applied before the leaves have been webbed together thus protecting the caterpillars. Spraying during flowering should be avoided. A second spray may be required in late summer to control the second generation of caterpillars.

Swift Moth Caterpillars

The larvae of two species, the ghost swift moth (*Hepialus humuli*) and the garden swift moth (*H. lupulina*) attack the roots of strawberries. Both moths are large and pale in colour, and are on the wing in June, July and August when the eggs are dropped in flight into herbage including strawberries. The young caterpillar which is white with a chestnut brown head feeds on the roots and can cause the collapse of plants in the same way as wingless weevils. Damage is more likely after grass or weedy crops.

CONTROL

Cultivation and weed control in the strawberry crop will do much to prevent infestation and the insecticidal treatments recommended against wingless weevils (see below) will also control swift moth caterpillars. Advisory Leaflet 160 also gives details.

Seed Beetles

Adult seed beetles (*Harpalus rufipes*) damage both ripe and unripe fruit by removing the seeds, causing superficial injury to the flesh. However, damage to strawberries by linnets is very often confused with that caused by beetles. It is important to distinguish between two types of injury so as to avoid unnecessary use of pesticides. The main difference is that linnets remove the seed cleanly, usually only from the upper surface of exposed fruit. Seed beetles injure the flesh around the seed and may attack any part of the fruit, particularly the underside next to the soil.

The adult beetles are a dull black colour with yellowish brown legs. They overwinter in sheltered places and move into the beds soon after strawing. Grass headlands, box stacks or nearby hedges and woodland provide ideal shelter for over-wintering beetles and should be avoided. A weed-free area around the fruiting beds will reduce the winter shelter. Other beetles (e.g. *Pterostichus* spp.) injure the fruit by eating large holes in the flesh; this is sometimes confused with slug or bird damage.

CONTROL

Where ground beetle attacks develop at the beginning of fruiting, damage can be considerably reduced by broadcasting a suitable pelleted bait over the rows, but not within seven days of picking. As an alternative a bran bait incorporating a suitable pesticide may be made: this remains effective for about seven days or longer in very dry conditions. A second application may be required. Rubber gloves should be used when handling the bait which should not be allowed to come into contact with the fruit.

Wingless Weevils

Several species may occur differing with locality. Vine weevil (*Otiorhynchus sulcatus*) and strawberry root weevil (*O. rugosostriatus*) are two commonly found species. The creamy-white, legless grubs feed on the root tips in the winter and spring, finally excavating deep tunnels in the base of the crown and so causing complete collapse of the plants. Attacked plants are usually isolated and collapse quite suddenly, so this damage need not be confused with verticillium wilt. The adults feed on the foliage at night during May and June. Leaf damage is unlikely to be important but serves to indicate that damage by the grubs may occur later. Crops adjoining woodland or mature black currant plantations are particularly at risk. Another weevil causing root damage is the red-legged weevil (*O. clavipes*), a locally serious pest in parts of the south west. There are two main periods of adult activity, the first in spring and again in mid-June and late August.

CONTROL

No satisfactory control measures can be taken against the well protected weevil grubs when the attack is first seen during flowering or fruiting. The aim must be to eradicate the infestation after fruiting to prevent further spread to adjoining strawberry crops. A pesticide drench to the crown of the plants and the surrounding collar of soil after cropping in mid-August for vine and strawberry weevil, or in early September for the red-legged weevil using 250–500 ml of dilute drench per plant is recommended. This will kill the young grubs hatching from eggs before they can penetrate deeply into the soil. Where land is known to be infested with these pests it should be treated three weeks before planting with a suitable pesticide well worked into the top 10 cm of soil. A non-tainting pesticide should be used on land to be cropped within eighteen months with potatoes, onions, carrots or other root vegetables for human consumption. Advisory Leaflet 57 *Wingless Weevils* also describes the pests and their control.

Strawberry Rhynchites (*Caenorhinusgermanicus*) and Strawberry Blossom Weevil (*Anthonomus rubi*)

Both species destroy strawberry blossoms and may occur on the plants at the same time. The adult weevils are small and seldom seen. The damage caused by each weevil differs. The greyish-black strawberry blossom weevil is generally the more common. Sometimes called the elephant beetle or needle bug it lays its eggs in the unopened flower buds and it then partially severs the individual flower stalks causing the flowers to wither. The strawberry rhynchites is a small blue-green weevil which punctures and lays its eggs in the main stem of the truss and young leaves, causing the whole truss to wilt and die.

CONTROL

Damage by strawberry blossom weevil can be reduced by spraying at the early green flower-bud stage or as soon as the first sign of damage is seen. A second application may be necessary ten days later. Where damage is noticed after the early flowers are open treatments should be applied in dull weather or late in the evening to reduce the danger to pollinating insects. To control rhynchites, the sprays should be applied before flowering in mid to late April.

Leafhoppers

Leafhoppers (*Aphrodes* spp. and *Euscelis* spp.) are the vectors of a disease that causes the green petal conditions in strawberry (page 99). The leafhoppers themselves do not damage the plants. They feed on a wide range of plants and are active for a considerable period, making chemical control measures difficult and of doubtful value.

Diseases of Major Importance

Grey Mould

Fruit rot or grey mould, caused by the fungus *Botrytis cinerea*, is a widespread and damaging disease of cultivated strawberry, it causes rotting of the green and ripening fruits, quickly reducing them to a soft mass covered with grey dusty spores of the fungus and, under humid weather conditions, can devastate the crop (Plate VII). The disease is also responsible for a reduction in yield due to *Botrytis*-induced abortion of flowers. Although there is some varietal variation in susceptibility this disease is mainly influenced by seasonal weather conditions.

Sources of *Botrytis* are very common and virtually impossible to eradicate. The fungus is an inhabitant of dead plant tissue from which it can attack living plants. Old dead strawberry leaves and leaf stalks provide excellent material for the fungus to live on and to produce spores, if given favourable weather conditions. The greatest danger comes from early infections of flower parts, such as the petals, sepals (calyx), stamens and receptacle by air-borne spores, or by contact with infected material. Infection at this stage may remain latent and then cause fruit rotting at harvest time. Fruit rotting also occurs by direct infection of ripening fruits by spores, or by contact of fruit with infected flower parts, or other rotting berries.

CONTROL

Good control of grey mould can only be achieved by ensuring a thorough protectant cover of fungicide during the blossoming period. Because of the long flowering period of the strawberry it is necessary to apply three or four sprays at the following times: (Plate II)

1. at white bud or first open flower
2. at early flowering; when first flowers open on most plants
3. at full blossom; when most of the flowers are open on most of the plants
4. at petal fall; when most petals have fallen on most of the plants

It is essential to commence spraying early in the flowering period, even in dry weather, as additional later sprays after blossom will not compensate for the omission of these early sprays, particularly if weather conditions favour development of the fungus. Later sprays will protect the late opening flowers but give problems of spray residues on developed fruits, especially when these are intended for canning or quick freezing, as well as the risk of damage by the passage of spray machinery.

A range of fungicides is approved for the control of strawberry *Botrytis*. In recent years, when some fungicides have been used exclusively for the control of this disease, strains of the fungus which are tolerant of these fungicides have been found. Where tolerant strains occur other fungicides must be used.

The efficacy of all fungicides depends largely on correct application; this is particularly so in the control of *Botrytis*. Thorough wetting of all plant parts, especially the flowers, is essential if good control is to be achieved; 2000 litre/ha (200 gal/acre) are necessary especially on established plants. The use of a ruffler bar (Plate III) to aid penetration of the spray into the crown of older plants is recommended. It is essential that the stipulated interval between last fungicide application and harvest is strictly adhered to, and, for crops intended for processing only fungicides approved by the processors should be used because of the risk of taint.

Attention to plantation hygiene can do much to lower the level of disease. Flailing off and burning over at the end of the fruiting season destroys the straw, leaf debris and large numbers of *Botrytis* resting bodies (sclerotia) which enable the fungus to survive the winter.

Red Core

Red core disease has been recorded in all the main strawberry-growing areas. The fungus, *Phytophthora fragariae,* lives in the soil and attacks the roots of strawberry plants causing them to rot. Infected plants are more often seen where drainage is poor and the incidence of red core is always worse following a wet autumn and winter. The effects of the disease are normally first noticed in late spring, or early summer, when diseased plants look as if they are suffering from drought; the plants are dwarfed and may wilt and die, particularly in dry weather. Their leaves are often dull bluish-green and the older leaves are sometimes tinged with brown or red. Occasional plants may wilt suddenly without previously showing any of these symptoms.

If the roots of an infected plant are examined in spring it will be found that the white feeding root tips have rotted away, so that only white main roots together with the old brown, rather woody, roots remain. The white roots may be brown and rotten for varying distances from their tips and have been described as 'rat's tail' roots. If these roots of an infected plant are cut lengthwise the characteristic red discoloration of the central tissue (stele) may be seen (Plate VII). This is the red core from which the disease takes its name. It is necessary to confirm red core infection by microscopic examination and even when characteristic red core symptoms in the roots are not visible plants can be infected.

The disease is spread from one area to another on infected runners. Contamination of land can also result from carrying infected soil on implements or other plants from field to field and by drainage and possibly irrigation water. Once the disease has been introduced into a field its development depends very largely on soil moisture conditions during autumn and spring. Under conditions of bad drainage, or high winter rainfall, the fungus readily invades strawberry roots to give the characteristic symptoms of the disease. In the infected tissues of the root the fungus forms thick-walled resting spores. When the infected roots rot away, these resting spores remain in the soil and apparently survive indefinitely. Once introduced into a field there is no known method of eradication.

CONTROL

It is very important to prevent the disease being introduced into clean land and every effort should be made to ensure that runners are free from infection by planting only certified runners. There is at present no cure for this disease and rotation of other crops, however lengthy, will not ensure that the disease has been eliminated. Much can be done to reduce losses from red core by attention to, and improvement of, field drainage. Growing plants on ridges appears to aid growth, presumably by improving surface drainage. Early planting, during August and late September, encourages root growth better able to tolerate infection before onset of winter than later planting. Replanting each year with strong healthy runners is common practice on fields where the disease is endemic. Plants grown under cloches and tunnels may crop satisfactorily although infected, as the protection keeps the roots drier in the late winter and spring.

Control of the disease by the use of resistant varieties is complicated by the existence of many strains or races of the fungus. Varieties differ in their field resistance to the disease. Cambridge Vigour, Redgauntlet and Gorella are tolerant of some, but not all, strains of the fungus. Three new varieties—Tantallon, Troubadour and Saladin all have field resistance. Cambridge Favourite is susceptible to all races of this fungus, but sometimes grows and continues cropping despite being infected, because of its high capacity for root regeneration which compensates for root losses due to the disease.

Recently a range of chemicals applied as a soil drench after planting, or as dips, have been found in trials to give some control of this disease, and some are being approved for use on both newly planted and established plantations. Their efficacy and persistence depends on soil type, methods of application and rainfall following treatment. Generally, treatment after planting has to be continued each autumn to prevent the fungus re-colonising the roots again. Those fungicides which are systemic (that is the chemical moves within the plant) may be easier to use. The latest information about which chemicals can be used each year is given in the books detailed on page 85. As all these control methods are expensive, avoiding infection with red core disease is still the best aim, coupled with good husbandry to encourage healthy plant root growth.

Under the *Red Core Disease of Strawberry Plants Order 1957* anyone who knows, or suspects, that plants growing on his land are infected must notify the Ministry of Agriculture, Fisheries and Food. The sale of infected plants or plants grown on contaminated land is prohibited. Such plants may not be planted except on land declared to be infected.
See also Advisory Leaflet 410 *Red Core of Strawberry*

Verticillium Wilt

This disease is caused by two fungi *Verticillium albo-atrum* and *V. dahliae* which live in the soil and under favourable conditions infect the plant causing it to wilt. The disease usually causes greatest loss in one-year-old plants. If plants are able to tolerate infection in the first year they usually survive to produce a crop in subsequent fruiting years. Yields from infected plants are appreciably below those obtained from healthy plants. The diseases can be introduced into clean soil by planting runners infected with verticillium wilt. Wilt is a less damaging disease than red core and its incidence fluctuates, depending on seasonal weather conditions; the disease is aggravated by soil compaction. Infection is often severe when strawberries follow other susceptible cultivated crops, especially potatoes.

This disease is generally seen in the summer but is difficult to diagnose correctly in the field. Infected plants wilt, especially during dry weather, the older leaves turn brown, shrivel and eventually the wilted plant may die. If the crown of an infected plant is cut transversely a brown discoloration may be visible in the wood tissue. Varieties differ in their susceptibility to wilt, Gorella, is very susceptible. Some of the new varieties are field resistant, see the chapter 'Varieties'.

CONTROL

Because of the problem of chemical control most growers try to avoid severe incidence of this disease by rotation and not planting on land known to be infected.

Chemical control by pre-planting soil injection is possible, but the material used is a noxious substance and may only be applied by registered contractors using specialist equipment. Some fungicides have reduced wilt severity when applied as a soil drench in the autumn and spring but there is a danger that these treatments may result in the selection of tolerant strains of *Botrytis* (page 92).

Diseases of Lesser Importance

Mildew

Mildew is caused by the fungus *Sphaerotheca humuli,* and is worse during dry periods. Leaves of attacked plants show characteristic dark red coloured blotches usually on the upper surface, later the margins curl inwards to expose the undersides which are covered with a white powdery mildew. Infected plants, showing these characteristic symptoms, appear to be suffering from drought. The fungus also covers the flowerstalks, flowers and fruit with a white powdery mildew.

CONTROL

Varieties differ in susceptibility to mildew infection—Cambridge Vigour, Gorella and Domanil, are all susceptible under field conditions. The disease can be controlled effectively by fungicides applied at early flowering, full blossom and again at petal fall. Some sprays sometimes cause damage to petals, and mixed sprays for control of both *Botrytis* and mildew may cause scorching of the foliage. For mildew control on runner beds of susceptible varieties spray at fourteen day intervals as necessary.

Leathery Fruit Rot and Leaf Blotch

Rotting of fruit can sometimes be attributed to attack by the fungus *Zythia fragariae.* The fungus can also cause a leaf spot and black sunken areas on the leaf stalks and stolons. The disease has been most damaging in wet summers when noticeable loss of fruit may occur in the susceptible varieties Redgauntlet and Cambridge Favourite.

Symptoms on the fruit are first seen as shrivelling of the calyx, which later curls upward. The disease progresses into the berry where it causes a slightly sunken, fairly firm, brown rot of the shoulder of the fruit. Tiny, flesh-coloured spore bodies may be seen in the rotted area under the calyx and, if the fruit surface is wet, a milky mass of spores may be produced. Attacked fruit soon become infected with *Botrytis* which very quickly masks infection with *Zythia.*

CONTROL

Limited observation studies have shown that early season sprays with zineb reduced the levels of *Zythia* attacking the foliage. The disease does not appear to be adequately controlled by the fungicides currently used to control *Botrytis.*

Crown Rot

In recent years, the disease caused by *Phytophthora cactorum* has been confirmed in this country on the varieties Cambridge Favourite, Redgauntlet, Gorella, Domanil, and particularly Tamella which is highly susceptible. The disease, once present in a field crop, can be devastating and it has also caused severe losses where strawberries are being propagated under glass or polythene. Spread and infection with the fungus is favoured by warm wet weather when soil temperatures exceed 10°C. In the field infection occurs mostly at flowering but may not show until much later. Cold stored runners are very susceptible immediately after planting. The youngest leaves on affected plants wilt, the symptoms spreading to the older leaves, followed by collapse, yellowing and death of the plant. At soil level necrotic areas develop in the crowns and infection can spread down into the roots. On infected soils, splashing of the fungus up onto leaves followed by infection is possibly more important than direct root infections; flowers and fruit can also be infected in this way.

In irrigated runner beds, infected mother plants can show variable symptoms. Infected runners may not show symptoms until favourable weather conditions occur for the fungus to become active. There is strong evidence that most infection is spread undetected in this way.

CONTROL

Apart from avoiding highly susceptible varieties like Tamella, there are no satisfactory control measures yet available.

Mucor and Rhizopus Fruit Rots

Fruit infection with these fungi results in softening followed by rotting of the fruit. *Mucor mucedo* sporulates on the fruit surface and is first seen as a grey mass of spore-containing bodies (sporangia) which turn black as they mature. The fungus spreads during storage by contact between healthy and infected berries. *Rhizopus stolonifera* and *R. sexualis* rapidly spread from one affected fruit onto others producing a mat of sporulating fungal mycelium.

Both fungi are more common on fruit harvested late in the season and where the incidence of rots due to grey mould has been reduced following fungicide treatment. The fungal growth of *Mucor mucedo*, *Rhizopus stolonifera* and *R. sexualis* can take place over a range of temperatures. The minimum temperatures at which growth occurs are 0°C for *Mucor mucedo* and 5°C for *Rhizopus stolonifera* and *R. sexualis*.

CONTROL

No fungicides yet available are effective against these fungi. Experience shows that fruit picked firm ripe and carefully handled is less susceptible than riper fruit. The fungi are troublesome on fruit preserved for later jam manufacture and particularly from plantations that have been under water stress.

Virus and Mycoplasma Diseases

Many different viruses affect strawberry and, although some cause only slight symptoms or none, the presence of more than one within a plant often results in a serious reduction of growth and crop. Symptoms are variable, depending on the viruses and virus strains involved, the strawberry variety and duration of infection. It is therefore often impossible to identify visually the virus or viruses present. Some strawberry viruses are spread from plant to plant by aphids, others by soil-inhabiting eelworms; mycoplasmas are transmitted by leafhoppers. Advisory Leaflet 530 describes and illustrates some of these diseases.

Aphid-transmitted Virus Diseases

Strawberry mottle, mild yellow edge, crinkle and vein chlorosis viruses are involved, all being spread by the strawberry aphid *Chaetosiphon fragaefolii*. Mottle is also transmitted, though less efficiently, by some other aphid species that infest strawberry. These viruses commonly occur together as complexes and the symptoms produced may be classified in four groups.

Yellow Edge is caused by infection with mild yellow edge virus together with mottle or occasionally crinkle virus. Some commercial varieties are sensitive to this disease. Typical symptoms, which appear in late summer and autumn, are stunting of the plant and marginal yellowing of the younger leaves. The yellow edges are not sharply defined but merge into the green parts of the leaf. The young leaflets are reduced in size and their margins tend to curl upwards, the leaf stalk is abnormally short and in severe cases the plant has a flattened, dwarfed appearance. Symptoms of yellow edge should not be confused with those of green petal disease (page 99).

Severe Crinkle results from mixed infection with crinkle virus and mottle virus. Young leaves of sensitive varieties develop numerous yellowish (chlorotic) spots which may become red or brown (necrotic). In the affected areas further expansion of the leaf is arrested, so that it becomes crinkled and distorted. In contrast with yellow edge the leaf stalks are not noticeably shortened and the plant retains its normal growth habit. Leaf symptoms are best seen in early summer but infected plants of many commercial varieties show no symptoms, although vigour and yield may be appreciably reduced. These virus symptoms can sometimes be confused with shallot aphid damage (page 86).

Mild Crinkle is caused by infection with mottle virus and vein chlorosis virus. In sensitive varieties the leaves show faint chlorotic spots, less numerous than with severe crinkle disease, and leaf distortion is absent or slight. Symptoms are more obvious in early summer and in the autumn but are seldom seen on most commercial varieties.

In recent years, Degeneration Complex, a virus disease causing stunting and severe leaf crinkling, has occurred in varieties that do not normally show crinkle symptoms. This is caused by a complex infection comprising, at least, strawberry mottle, mild yellow edge and crinkle viruses.

CONTROL

Many commercial varieties are tolerant of these aphid-transmitted virus diseases, especially mild or severe crinkle. The certification schemes providing healthy planting material, and efficient aphid control have reduced these diseases. Fruiting plants should be treated with a systemic insecticide in spring to give a good control of aphids and reduce the risk of viruses being spread (page 86), If isolated infected plants are seen they should be dug up and burnt. Runner beds must be sprayed more frequently to ensure adequate protection.

Eelworm-transmitted Virus Diseases

Strawberries are affected by four diseases of this type—arabis mosaic, strawberry latent ringspot, raspberry ringspot and tomato black ring viruses. They are transmitted to healthy strawberries by root-feeding eelworms. In England and Wales, arabis mosaic virus is the only one of economic importance; but in Scotland mixed outbreaks of raspberry ringspot and tomato black ring viruses often occur.

All these viruses affect a wide range of plants, including several of economic importance; also many common weeds and hedgerow species which are often symptomless. Arabis mosaic and strawberry latent ringspot viruses are transmitted by the eelworm *Xiphinema diversicaudatum* and the other two by species of *Longidorus*. The range of symptoms shown by strawberry varieties is very wide, but these viruses generally cause stunting of the plant and yellowish patterns on the foliage. Symptoms are most pronounced in late spring and to a lesser extent in autumn, but although leaf patterns tend to fade during the summer, infected plants remain stunted. Arabis mosaic virus usually produces yellow spots on the leaves (Plate VII) In many varieties these spots become bright red and leaf crinkling occurs in most varieties. Cambridge Favourite is very sensitive to this virus, developing distinctive yellow blotches more quickly than other varieties. Conspicuous leaf symptoms produced in Cambridge Vigour are usually transient; but there are strains of the virus that give symptoms in this variety almost as severe as those seen in Cambridge Favourite.

When infection arises through planting into soil infested by virus-carrying eelworms, these diseases occur in well defined patches which usually extend slowly within the crop, sometimes from an existing or former hedgerow site. If runners were infected before planting, diseased plants will generally be found scattered amongst healthy ones. Planting strawberries on land that has not previously grown the crop is no guarantee that virus infection will not occur.

CONTROL

Because soil treatment before planting is expensive and symptoms of arabis mosaic virus can be confused with aphid-borne severe crinkle, it is important to have suspected eelworm-transmitted infection checked. Not every eelworm population carries virus infection but this is likely where there has been previous trouble. Growing crops which are not susceptible to the viruses (such as cereals) for two or more years between strawberry crops will greatly reduce eelworm numbers, provided that weeds are rigidly controlled throughout the year. *X. diversicaudatum* will not die out during this period, but will lose arabis mosaic virus if denied further access to susceptible plants. If isolated plants with symptoms of arabis mosaic virus appear, they should be removed with the surrounding soil and burnt. Pre-planting soil fumigation is difficult (page 89).
See also Advisory Leaflet 530 *Virus and Mycoplasma Diseases of Strawberry*.

Leafhopper-transmitted Mycoplasma Diseases

Green petal and bronze leaf wilt were formerly attributed to viruses but are now known to be caused by very minute organisms called mycoplasmas, which are spread by leafhoppers, mainly *Aphrodes bicinctus*. Clover, several ornamentals and some weeds are also susceptible and can constitute additional sources of infection for strawberry plants. Symptoms in the strawberry usually appear in the year following infection and are more prevalent after a dry summer, which increases leafhopper activity and hence dissemination of the mycoplasmas. Because the spread of infection in strawberry fields is normally slow, these diseases are less important than those caused by aphid-borne viruses.

In spring, plants infected with green petal may be stunted and distorted, their young leaves chlorotic, small, and often asymmetrical, with leaflets of unequal size; but symptoms develop slowly so that older plants may have both affected and healthy crowns. The petals are smaller than normal and sometimes greenish in colour, but the main symptom is failure of fruits to swell, these remaining 'pippy' and leathery. After flowering time the mature foliage becomes reddish or bronze in colour. Plants developing symptoms early in the season usually wilt and die during the summer months; others produce only very small leaves during the autumn and gradually die. All varieties are sensitive to the disease but some seldom show greening of petals.

Bronze leaf wilt is induced by mycoplasma similar to that of green petal but there are no flower symptoms. Older leaves turn a marked bronze colour and infected plants wilt and die more rapidly than those with green petal, often in full fruit. The varieties Cambridge Favourite and Talisman are particularly sensitive to this disease.

CONTROL

Plants with symptoms should be avoided for propagation, any found in runner beds with all their runners being removed and destroyed. It is worth while removing isolated infected plants from young fruiting plantations. Control of leafhopper vectors is difficult because these may be sporadically active over a long period (page 99). Wide, clean headlands may help to reduce the spread of mycoplasma infection into runner beds.

Damage from Vermin

Birds can cause serious damage to ripening fruit and may be particularly troublesome on small areas, or when damaging the early ripening berries. Main offenders are blackbirds, thrushes and linnets. Various banger guns may give some control but care must be taken to ensure that County byelaws permit their use. Other scaring devices include artificial hawks suspended from helium-filled balloons and various moving, or noise-making, equipment. All are of limited value as birds quickly become accustomed to them. Once the crop is in full harvest with pickers around all day, the loss due to birds is usually less severe. Strawberries under tunnel houses and cloches may be protected by closing the ends with wire, or nylon netting, which keep out birds but permits ventilation.

Rabbits and particularly hares, will sometimes attack young strawberry plants during the autumn and winter, eating the leaves but generally leaving the crown.

They may be particularly troublesome on runner beds. Where severe damage is expected, these areas should be temporarily fenced with rabbit-proof netting.

Physiological Leaf Disorders

Tipburn of leaves is a physiological disorder, not entirely understood but apparently caused by a local deficiency of calcium in emerging leaves, resulting in the death of leaf tips and sometimes of much of the leaf blades. Subsequently the dead tip restricts expansion of the growing leaf blade and the characteristic leaf distortion results, from which the individual leaf never recovers. Varieties differ in susceptibility. Cambridge Favourite is very susceptible but not Gorella.

This condition should not be confused with spray or wind damage which usually occurs over the whole plant.

Although aggravated by calcium deficiency at the roots, tipburn is usually brought on indirectly by high humidity, which perhaps restricts the movement of minerals into the emerging leaves or by excessive amounts of potassium, ammonium or other ions in the soil solution, or simply by drought. Tipburn is seldom severely damaging outdoors, and corrective measures are not usually necessary.

Emerging trusses may be damaged by tipburn in glasshouse culture, and the consequences can then be serious. Avoiding excessive potassium, ammonium or urea feeds and avoiding high humidity are the best safeguards indoors.

Spray Machinery for Pest and Disease Control

Many pests and most diseases can only satisfactorily be controlled in the strawberry crop by the use of machinery that will apply high volume (HV) sprays 1000 to 2000 l/ha (90–180 gal/acre). Sprayers used for ground crop spraying may be satisfactory for application of herbicides but are unsuitable for pesticides.

The choice of spray machinery for the strawberry crop may depend on what other crops are grown on the holding and the sprayer to work in all these crops may not necessarily be the best for strawberry plantations.

The strawberry is a very difficult target to cover adequately with spray materials as it is thick and compact in growth, unlike the top and bush fruits. Fungicides for the control of rots have to be applied before damage occurs and these protectant sprays need to cover all parts of the flowers and early fruits, so this spraying is particularly difficult. Chemicals for pest control are sometimes systemic, that is they can move within the plant tissue, particularly to the growing parts of the plant and the pests themselves may move about and so come into contact with the chemical deposits. But both aphids in the crowns of the plants in the spring and red spider mites, need very thorough wetting of the plants to ensure good control.

Most growers use tractor mounted hydraulic sprayers with pumps of high output so that at slow tractor speeds the required HV sprays per area can be applied. The tractor PTO rating must be adequate both in power and rpm to run the sprayer. Special strawberry spray booms are available carrying adjustable HV nozzles. The nozzles can be angled to spray one on each side of the plant row and one from above. To improve penetration of the sprays into the centre of the plants a ruffler bar or chain is often fitted (Plate III); this bends the leaves over and aids the spraying of the centre of the plants. The chain type is better as it is more easily adjusted and follows the contours of the ground. Obviously ruffling is more important on thick older plants and dense matted rows than on young, spaced plants. Some strawberry booms only have one nozzle spraying the row centre: this may not always provide adequate cover. All booms should have nozzles that are readily adjustable in position to cover different row widths. Calibration of sprayers is essential (page 55) to ensure correct application.

Air blast sprayers used for top fruit are not suitable for spraying strawberry plants as most of the spray is blown out well above ground; these machines are not designed for HV application to ground crops (see Appendix I for ADAS leaflets describing sprayer types).

Research is being done on the use of controlled droplet application (CDA) and ultra low volume spraying (ULV) on many crops in an effort to reduce the quantities both of water and pesticides used. At the date of publication of this book none of these had been cleared or approved for use on strawberries.

Growers of small areas of strawberries may need to use hand lances from tractor sprayers which can work from the headlands. This method, or use of knapsack sprayers, may be necessary for spraying in film plastic houses if tractors cannot proceed. Crops in low continuous tunnels can be satisfactorily sprayed by the standard strawberry booms, but the sprayer may have to be mounted on high row wheels to clear the hoops, and if the soil is wet, less quantity of spray carried each time to avoid soil rutting.

Some pesticides can be applied as soil acting granules; useful materials are available for the control of aphids, and there is a range of specialist tractor mounted and hand applicators, which are widely used in vegetable production, available for these materials.

With all sprayers of whatever kind efficient maintenance and accurate calibration is essential in order that the correct dosage can be applied. Sprayers should be serviced each winter and all nozzles renewed. ADAS surveys have shown that nozzle wear—and blockage—are one of the most common faults in inefficient spraying.

Spraying of pesticides should always be carried out under the best possible conditions of wind, temperature and humidity to avoid drift, ensure good penetration and avoid phytotoxicity. It is important that the sprayer has an acceptable agitator to keep the chemicals properly mixed at all times and this should be frequently checked.

The manufacturers' instructions should always be carefully followed and materials accurately measured, or weighed out, and correctly mixed. If more than one chemical is being applied check that the mixture is both compatible and safe. Many materials require a minimum interval between the last spray and expected harvest date. Strawberries intended for processing should not be sprayed with pesticides that cause taint; the processing firm, or agent, will supply a list of acceptable chemicals.

Employer, employee and the self-employed must follow certain regulations regarding protective clothing, recording, disposal of waste sprays and containers and other actions when using chemicals (Appendix II).

Many sprays, particularly certain herbicides, can contaminate spray machinery, especially leather or rubber washers and cause damage later if the machine is used without adequate cleansing. After all spray operations pump out surplus material into an approved soakaway area, fill up with clean water and pump through, adding a suitable wetting agent if contaminating materials have been used, refill, remove nozzle including stop ends to the booms and pump out.

A final reminder to all growers is that no chemical is better than its application. Many complaints about lack of control of pests or diseases on the strawberry have been shown to be due to wrong timing, or inadequate application or, worst of all, both these common causes of failure combined.

Economics

The strawberry is the cheapest of all fruit crops to establish and bring into production. The average arable farm will possess generally the necessary tractors, cultivating and fertiliser machinery and probably also a planting machine that will plant strawberries. A ground crop sprayer suitable, with careful cleaning, for strawberry herbicides is now present on most farms together with a flailer, or mower, for defoliation after harvest if necessary and possibly a rotary cultivator. The only specialised equipment that may need purchase is a suitable, tractor mounted, hydraulic HV sprayer together with special strawberry booms and adjustable nozzles. A second-hand sprayer may be available that will be suitable if fitted with the appropriate boom(s).

For harvesting on a large scale pallet transporters on tractors that can quickly run the stacked, full market trays into the cool holding area will suffice, or covered trailers may be used. If PYO is adopted as the sales method capital expenditure will be necessary for suitable pay out shed(s), weighing scale(s), fencing and car parking. Improvement of access points and inner farm roads may also be required.

To plant one hectare of outdoor strawberries the following materials and labour are required. The approximate costs given here are for 1979 and based on ADAS costings in Eastern England.

Commodity	Quantity	Cost (£)
Plants	25 000	700
Fertiliser	—	80
Herbicides	—	20
Insecticides	—	90

Labour	Labour hours: Regular	Casual
Subsoiling, ploughing		
Cultivations & fertilisers	25	–
Marking out, planting	26	48
Herbicide application	6	–
Pest control	2	–
Hand weeding	4	30
Runner control	6	–
Gapping up	–	8

At 1979 rates these labour hours cost about £270. An allowance of £600 per hectare is needed for overheads (of the farm fixed costs). These total costs could be written off over four crops at ten per cent interest to give an annual charge for establishment of the crop.

In each production year there are the following costs:

	1979 Costs
	£
Fertiliser	40
Herbicides	80
Insecticides	25
Fungicides	100
Straw	40

	Labour Hours	
Labour	Regular	Casual
Fertiliser application	2	–
Pesticide applications	12	–
Strawing	8	80
Hand Weeding	4	25
Cleaning after harvest	8	5
Herbicide application	6	–

At 1979 rates these labour hours would cost £270. Harvesting is the main labour cost and obviously depends on the yield per hectare. Many strawberries in the main production areas are picked by piecework (1979) at 18p/kg or £170 per tonne. Supervision, £30/tonne, and making up boxes and transport of pickers costing £30/tonne gives a total harvesting cost of about £230/tonne.

In addition to the labour harvesting costs there are the costs of market containers—punnets 225/g £50/tonne, market trays £35 and transport £40/tonne, a total harvesting and marketing cost of about £355/tonne, plus the normal market commission.

For processing a slightly lower piecework rate is paid and there is no cost for non-returnable containers. Picking, supervision and transport of pickers and weighing cost £200/tonne.

Costs of harvesting for PYO vary widely and it is difficult to give any exact figures. Containers (mushroom baskets) cost £40/tonne for strawberries and allowance must be made for supervising and check out staff and advertising.

It will be apparent from the foregoing that good yields per hectare are necessary to spread these heavy establishment, overheads and growing costs. This is particularly so with protected cropping where closer density, the use of FYM, much hand labour, plastic covers only lasting one year and more careful picking give 1979 estimated costs per hectare for low continuous tunnels of £4700 and harvesting costs of £500/tonne. Under film plastic tunnels estimated costs are even higher at £7500 for labour and materials and the £500/tonne for marketing.

These costs can only be a rough guide but they do provide a yardstick to measure individual grower efficiency, provided allowance is made for any inflation adjustment necessary.

Appendix I

Reference to other MAFF publications relevant to strawberry growing and kindred subjects

Advisory Leaflets

These are obtainable from advisers of the Agricultural Development and Advisory Service on request, or in conjunction with advisory enquiries. They are revised more frequently than this Book and so can contain more up-to-date information. Leaflets dealing with strawberries are:

EEC Standards for Fresh Strawberries	
Fruit Growers Guide to the use of Chemical Sprays	HFG 21
Direct Farm Sales to the Public	
Field Crop Sprayers	ML 2
Horticultural Sprayers for small areas	Booklet 2070
Windbreaks	HG 21
Early Strawberry Production from Plastic Tunnels and Houses	STL 78
Promising Varieties of Soft Fruits	HSG 24
Cold Storage of Strawberry Runners and their use	STL 54
Lime and Fertiliser Recommendations No. 3 Fruit and Hops	GF 23
Chemical Weed Control in Strawberries	HSF 21 (STL 23)
Red Core of Strawberry	AL 410
Virus and Mycoplasma Diseases of Strawberry	AL 530
Eelworms on Strawberry	AL 414
Strawberry Mite	HSD 51 (AL 584)
Swift Moths	AL 160
Red Spider on Outdoor Crops	HGD 53 (AL 226)
Wingless Weevils	AL 57

Advisory Leaflets are available from MAFF Publications, Tolcarne Drive, Pinner, Middlesex HA5 2DT

Books and Bulletins

These are obtainable through any bookseller or direct from HMSO, PO Box 569, London SE1

Fertiliser Recommendations for Agricultural and Horticultural Crops	GF 1
Soils and Manures for Fruit	Bulletin 107

Irrigation	Bulletin 138
Water for Irrigation	Bulletin 202
Home Preservation of Fruit and Vegetables	Bulletin 21
Refrigerated Storage of Fruit and Vegetables	RB 234
Approved Products for Farmers and Growers	Annual Publication (February)
Horticulture in Britain Part 2 Fruit and Flowers	
Bush Fruits	Bulletin 4
Cane Fruits	Bulletin 156

Proprietary products based on chemicals used for pest, disease and weed control can be officially approved under the Agricultural Chemicals Approval Scheme. It is strongly recommended that approved products should be used. Approval is indicated on the containers by the mark shown here. The booklet *Approved Products for Farmers and Growers* is available as a priced publication from HMSO, PO Box 569, London SE1 9NH or through booksellers.

Appendix II

Care with Chemicals

Whenever chemicals are used, FOLLOW THE INSTRUCTION ON THE LABEL.

Read: 0 "Some advice on the safe use of agricultural chemicals" at the beginning of *Approved Products for Farmers and Growers.*

Consult: 0 Chapters in *Approved Products for Farmers and Growers* on
0 Health and Safety (Agriculture) (Poisonous Substances) Regulations
0 Chemicals subject to the Poisons rules
0 First Aid Measures
0 Application of Agricultural Chemicals

PROTECTION OF STAFF

The Health and Safety at Work Act 1974 imposes obligations on employers, the self-employed and employees that work with any chemicals

Read: 0 *The Safe Use of Poisonous Chemicals on the Farm* (available from offices of the Health and Safety Executives). Paraquat is subject to the Pharmacy and Poisons Act 1933. It is only available to bona fide farmers and growers who have to sign the poisons register on purchase.

PROTECTION OF CONSUMERS

Allow at least the minimum intervals, as stated on the label, between the last application of chemicals and harvesting.

PROTECTION OF THE ENVIRONMENT

To safeguard bees, do not spray strawberries with insecticides during blossoming.

Dispose of empty containers safely.

Read: 0 *Code of Practice for the disposal of Unwanted Pesticides and Containers on Farms and Holdings*

Do not contaminate ponds, ditches or waterways.

Store new and part used containers in a secure place.

Appendix III

Tariffs on Strawberry Imports to the UK (July 1979)

Country	Tariff	
European Community Member States	Free	All the year round
Greece	Free	All the year round
Countries other than those shown below	Common Customs Tarrif (CCT)	*viz* 16 per cent, with a minimum of 3 u.a.* per 100 kg net, for the period 1 May–31 July and 14 per cent for the period 1 August–30 April
Cyprus Malta Israel Algeria, Morocco, Tunisia (Magreb Countries)	CCT – the above rates applied as appropriate to the period 1 April–31 October	5.6 per cent for the period 1 November–31 March
Portugal	CCT – the above rates applied as appropriate to the period 1 March–30 September	11.9 per cent for the period 1 October–last day of February

*3 u.a. (units of account) per 100 kg is equivalent to £20.08 per tonne.

Printed in England for Her Majesty's Stationery Office
by McCorquodale Printers, Ltd., London
Dd597192 K26 9/79